Quantum Many-Body Systems in One Dimension

SERIES ON ADVANCES IN STATISTICAL MECHANICS

Editor-in-Chief: M. Rasetti

Published

Forthcoming

Series on Advances in Statistical Mechanics – Volume 12

Quantum Many-Body Systems in One Dimension

Zachary N. C. Ha

Institute for Advanced Study, USA

 **World Scientific**
Singapore • New Jersey • London • Hong Kong

Published by

World Scientific Publishing Co. Pte. Ltd.

P O Box 128, Farrer Road, Singapore 912805

USA office: Suite 1B, 1060 Main Street, River Edge, NJ 07661

UK office: 57 Shelton Street, Covent Garden, London WC2H 9HE

British Library Cataloguing-in-Publication Data
A catalogue record for this book is available from the British Library.

First published 1996
First reprint 1998

ISBN 981-02-2275-0

Printed in Singapore.

To Seojin,
Parents,
and
God.

Preface

This book is an outgrowth of my Ph.D thesis at Princeton University, and major part of the work is guided, inspired by and collaborated with Professor Duncan Haldane. The book is organized roughly in chronological order which the subject was actually developed, and it follows a pedagogical guiding principle, "learn by going through examples."

A common thread that connects various models in this book is so called "strings" which represent the structure of excitation spectra and can be considered as quantum solitonic excitations with in general internal quantum degrees of freedom. The first chapter is an introduction to 1D systems and the following two chapters discuss some features of the representative models with nearest neighbor interaction, namely, the Heisenberg spin chain and the Hubbard model. The Bethe-ansatz method is discussed in detail and for the Hubbard model a novel strong-coupling thermodynamic expansion (referred to as "λ-expansion") is developed. Next four chapters concentrate on interesting properties of models with long-range interaction and discuss striking similarities and differences with the Bethe-ansatz solvable class. Perhaps the most striking feature of the long-range interaction model is that the elementary excitations form ideal gas that supports 1D fractional statistics and, therefore, some dynamical correlation functions can be calculated exactly. This interesting property is discussed in Chapter 7. Some concluding remarks are made in the last chapter.

Various people have contributed for the publication of this book. First, I would like to express my deep gratitude to Duncan who influenced me in scientifically most profound way. I would also like to thank the following people who have influenced me in positive ways: P. W. Anderson, G. Baskaran, Y. M. Cho, A. Georges, R. Narayanan, J. Talstra, F. Wilczek, etc. Finally, I would like to give special thanks to my parents and my wife Seojin whose constant support and love are beyond compare.

Zachary Ha
Princeton, NJ
May 1996

Contents

Chapter 1

Introduction

Everything has beauty, but not everyone sees it.
Confucius

The intimate interaction between natural science and technology (namely, the use of technological innovations in studying fundamental laws of nature and the use of scientific guiding principles in developing new technology) in the past few centuries have been extremely productive to say the least. The discoveries of atoms, electromagnetism, quantum mechanics, subatomic forces/particles, galaxies, quasars, pulsars, microwave background radiation, superconductivity, and practically all other fundamental and exotic natural phenomena in this and last centuries would not have been possible without the aids of technology, and by the same token the inventions of transistors, telecommunication devices, lasers, computers, and so forth that have forever change the world we live in would unlikely have been achieved without the guiding principles of physics and other branches of science. With science comes the understanding, and with technology comes the practical use of the scientific knowledge and new tools for further scientific research and so on. This feedback loop of influence between science and technology as depicted in Fig. 1.1 has had phenomenal effects on each other and continues to grow in intensity.

The interaction has been so intense in the past couple of decades that the technology rather than Nature herself is beginning to supply "fundamental" problems for some branches of science including the condensed matter physics. The modern technology allows some control over designing and manufacturing of physical systems at microscopic scales. Some of the best known example of these pre-processed or semi-artificial systems[1] in condensed matter physics include the high temperature superconducting cuprates, fractional quantum Hall (FQH) systems, quasi-one-dimensional organic superconductors, quantum dots, quantum wires, one-dimensional spin chains, etc. And, some of these physical systems exhibit novel properties that require fundamentally new physical concepts. The peculiarities of the normal state of the cuprates above the superconducting temperature, for example, are not well understood in the

---

[1] I call these semi-artificial since there is a certain length scale (usually larger than the atomic scale) below which the effect of human manipulation is minimal.

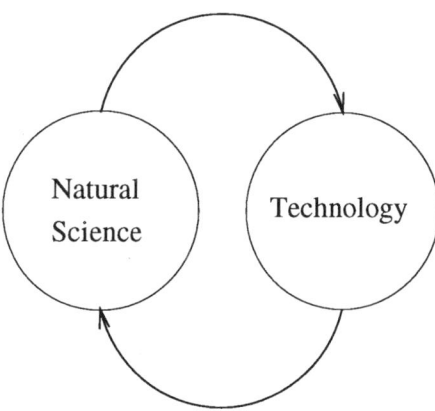

Fig. 1.1. Feedback loop of influence.

context of the usual Landau Fermi liquid theory and probably requires fundamentally new concept for proper and satisfactory understanding [1]. Signatures of novel phenomena such as fractional statistics, "Luttinger liquid" and Haldane gap are all observed in semi-artificial systems such as the quantum Hall bars, 1D organic metals at low enough temperatures and 1D integer spin chains. In any case the advances in science and technology have enlarged the meaning of what is fundamental and what is real and physical. A pictorial illustration of this intense interaction between science and technology giving birth to a new field of study is shown in Fig. 1.2.

One of the goals of this science/technology is to have more control over substances at smallest possible scales and to design and manufacture useful substances. What can be achieved with this science/technology is limited only by human imagination; and, consequently, the meaning of "unphysical" has to be modified. One-dimensional (1D) systems, for example, were once considered to be unphysical, but now 1D quantum systems can be made in laboratories. In this book I theoretically explore this 1D quantum world, a new emerging frontier of science/technology.

One advantage in studying 1D quantum systems is that due to the highly restrictive spatial degrees of freedom there exist many exactly solvable models. The spatial restriction itself, however, introduces large quantum fluctuations and prevents the use of known techniques that generally work for higher dimensional systems. Therefore, one has to be careful in studying the general features of 1D systems. Since a good way for a child to develop cognitive skills and intuition is to play with many toys, I will play with many 1D toy models throughout this book to help the reader build intuition on the novel features of 1D physics.

In this chapter I introduce the following three classes of exactly solvable 1D many-body models,

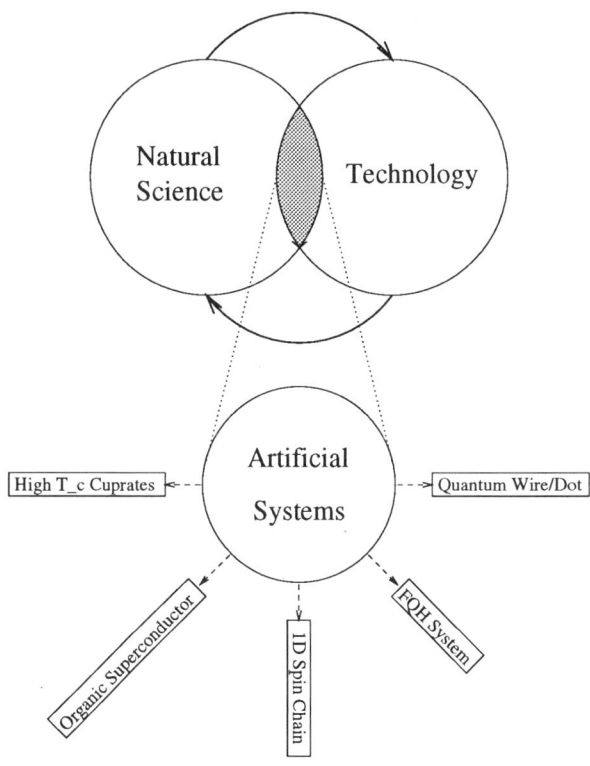

Fig. 1.2. Merging of science and technology and birth of new field.

- A) models solved by Bethe-ansatz,

- B) models with long-range interaction,

- C) linearized models.

The models in class A have characteristic nearest-neighbor exchange and appear to be very different from those in class B. However, there are many striking similarities in the two classes of models. Furthermore, the zero temperature behaviors of models in both classes A and B can be described by models in class C. More generally, all of these models (with the exception of models exhibiting low energy gaps) in the limit of zero temperature belong to an universality class called "Luttinger liquid" [103, 80, 37, 38].

The pioneering works in Luttinger liquid theory have exposed some important general low-temperature features of 1D liquid such as the spin-charge separation, but still there are features missing from the general analysis. For example, the elementary excitations of the spin-half Heisenberg chain have been identified to be the unbound pairs of spinons with $S = 1/2$ by Faddeev and Takhtajan [20] using the exact Bethe-ansatz solutions and, recently, the spinons have been experimentally observed [102]. One could not have predicted this special feature of 1D spin system from the general Luttinger liquid theory. While some features of specific models are general many are not and, therefore, the best strategy in understanding 1D system is to combine the features of the general theory with those of many solvable models.

1.1 Models solved by Bethe-ansatz

All of the models solved by the Bethe-ansatz share a common feature that their interactions are given by either the nearest-neighbor exchange or the Dirac-delta function depending on whether the systems are on lattice or in continuum space. The very first model that started this branch of physics is the isotropic spin-half chain with nearest-neighbor exchange. This was first solved by H. A. Bethe in 1930s [9] using a substitution method, and thus the name Bethe-ansatz. More detailed study of this chain is given in the following chapter, and only brief description of the model is given in this section.

The hamiltonian for the spin chain is given by

$$H = \sum_{\langle i,j \rangle} J_{ij} \vec{S}_i \cdot \vec{S}_j, \tag{1.1}$$

where $\vec{S}_i$ is the $S = 1/2$ local spin operator at ith site and the sum over all the distinct spin pairs. The model Bethe solved is a special case $J_{ij} = J\delta_{i\pm1,j}$, i.e. the nearest-neighbor spin exchange, and the famous ansatz he used to diagonalize the corresponding hamiltonian is given by

$$\sum_{P} A_P e^{i \sum_j k_{P_j} n_j}, \tag{1.2}$$

where the sum is over all possible permutations of a set of distinct wave numbers $\{k_1, k_2, \ldots, k_M\}$ corresponding to M down spins. Since the total S_z commutes with the hamiltonian the total numbers of up and down spins are conserved; therefore, the spin system is equivalent to a system of hard-core bosonic particles (down spins) moving in the background of holes (up spins).[2] The eigen-energy and total (crystal-) momentum are determined, respectively, by

$$E \;=\; J \sum_j \cos(k_j), \tag{1.3}$$

$$P \;=\; \sum_j k_j. \tag{1.4}$$

(For the lattice case the pseudo-momenta k_j are restricted to $0 < k_j < 2\pi$.) Hence, the problem of finding the eigen-spectra is reduced to finding all possible sets of pseudo-momenta $\{k_j\}$.

An intuitive reason why the plane wavefunction (1.2) yields the solution is that in one-dimensional system with a pair-wise interaction the magnitudes of momenta of two colliding particles are conserved owing to the two conservation laws for energy and (crystal-) momentum.[3] Hence, the two particles can pass through each other and thus preserve their original momenta or exchange their momenta (up to the reciprocal lattice vector for the lattice system). It is therefore at least reasonable for a M-particle system (or the corresponding spin system) to admit the wavefunction given in Eq. (1.2) which is just a weighed sum of the plain wavefunctions corresponding to all possible permutations of the given set of momenta.

The isotropic spin hamiltonian can be generalized to anisotropic [54, 87], to $SU(2)$ higher spin[4] [2], to $SU(n)$ "spin" or more generally to supersymmetric $SU(n|m)$ case in which the system admits n kinds of spins for bosons and m kinds for fermions. The spin degrees of freedom in the $SU(n|m)$ case are represented by the graded Lie group [15]. The $SU(1|2)$ case is the well-known supersymmetric t-J model whose hamiltonian is given by

$$H = -\sum_{\langle i,j \rangle, \sigma} t_{ij} \left(c_{i,\sigma}^\dagger c_{j,\sigma} + c_{j,\sigma}^\dagger c_{i,\sigma} \right) + 2 \sum_{\langle i,j \rangle} J_{ij} \vec{S}_i \cdot \vec{S}_j, \tag{1.5}$$

where $c_i(c_i^\dagger)$ is the fermion destruction (creation) operator and $\vec{S}_i$ the spin operator and $t_{ij} = J_{ij} = t\delta_{i\pm 1,j}$. The first sum describes the hopping and the second the nearest-neighbor spin-spin interaction.

Another well-known model that can be solved by the Bethe-ansatz method [77] is called the Hubbard model whose hamiltonian is given by

$$H = -t \sum_{\langle i,j \rangle, \sigma} \left(c_{i,\sigma}^\dagger c_{j,\sigma} + c_{j,\sigma}^\dagger c_{i,\sigma} \right) + U \sum_i n_{i\uparrow} n_{i\downarrow}, \tag{1.6}$$

---

[2]Of course one can also take the up spins as particles and down spins as holes.

[3]The crystal momentum of a system on lattice is conserved up to the reciprocal lattice vector.

[4]In this case the integrable hamiltonian is not simply given by Eq. (1.1) but rather a polynomial of the spin dot-product $\vec{S}_i \cdot \vec{S}_j$.

where U parameterizes the on-site repulsion or attraction. There has been a surge of interest in this model in recent years because a correct description of the copper oxide planes in the High T_c superconductors is believed to be given by the 2D version of this model which has not been solved. For now, therefore, people resort to the 1D version and find out as much information as possible. A detailed treatment of this model is given in Chapter 3.

A model that predates the Hubbard model and whose solution gave birth to the so called Yang-Baxter equations is actually a continuum system of $SU(2)$ fermions interacting with the Dirac-delta function. The hamiltonian is given by

$$H = -\frac{\hbar^2}{2m} \sum_j \frac{\partial^2}{\partial^2 x_j} + c \sum_{i<j} \delta(x_i - x_j). \tag{1.7}$$

The solution to this model is strikingly similar to that of the Hubbard model.

The list of models given in this section is by no means complete, but it consists of most commonly used and referred to. Even though these models are *exactly* solvable they are not *fully* solvable. The wavefunctions are notorious complicated that any analytic calculations such as the dynamical correlation functions using the wavefunctions are almost impossible. This feature is one of the main drawbacks of this class of models.

1.2   Models with long-range interaction

1.2.1   Inverse-square exchange models

Recently, there have been much interests in a class of models with the interaction that falls off as the inverse-square of the distance between a pair of particles or spins. This type of models dates back to Moser [86] (classical systems) and Calogero [11] and to Sutherland [95]. The quantum hamiltonian in units of $\hbar^2/2m$ for Calogero-Sutherland model (CSM) is given by

$$H = -\sum_{j=1}^{N} \frac{\partial^2}{\partial^2 x_j} + \sum_{j<l} \frac{2\lambda(\lambda - 1)}{d^2(x_j - x_l)}, \tag{1.8}$$

where $d(x) = (L/\pi)\sin(\pi x/L)$ and L is the circumference. The ground state of this model is interestingly given by a 1D version of the famous Laughlin's wavefunction

$$\Psi = \prod_{j<l}(z_j - z_l)^{\lambda} \prod_k z_k^{J_0}, \tag{1.9}$$

where $z_j = \exp(i2\pi x_j/L)$ and $J_0 = -\lambda(N - 1)/2$.

The Calogero-Sutherland model has recently been identified with ideal fractional statistics and is further related to many to other branches of mathematics and physics as depicted in Fig. 1.3. Among them are Selberg integrals [21, 22], W^∞ algebra [52], morphology of vicinal crystal surfaces [105], edge states of quantum Hall droplet

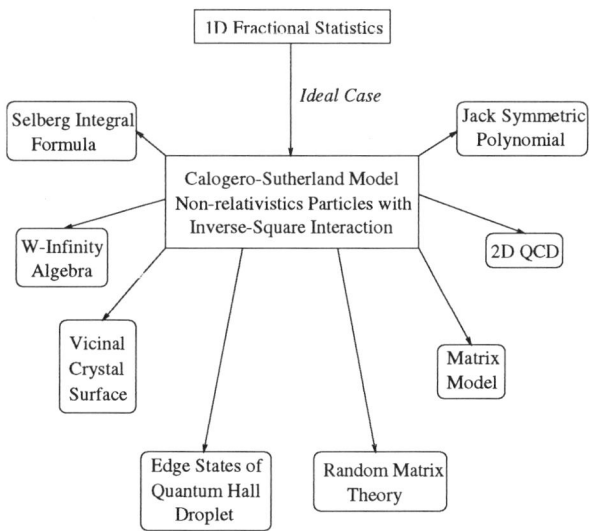

Fig. 1.3. Calogero-Sutherland Model and related subjects from W^∞ algebra to 2D QCD to fractional statistics.

[48, 35], random matrix theory [17, 84], Jack symmetric polynomials [56], etc. Some of the related subjects are discussed later in this book and, especially, the aspect of fractional statistics is explored in detail in Chapter 7.

A lattice version of CSM is independently introduced by Haldane and Shastry [40], thus, called Haldane-Shastry model (HSM) whose hamiltonian is given by

$$H = J \sum_{i>j} \frac{P_{ij}}{d(i-j)^2}, \qquad (1.10)$$

where $d(n) = (N/\pi)|\sin(\pi n/N)|$ which corresponds to the chord distance between the pair of spins separated by n sites along the arc of a circle with N equally spaced sites and P_{ij} is the spin exchange operator, e.g. $P_{ij} = (\vec{S}_i \cdot \vec{S}_j + 1/2)/2$ for $SU(2)$ case. The exact ground state and its current excitations for the hamiltonian (1.10) are given by the following wavefunction

$$|\Psi\rangle_{N,M} = \sum_{\{n_i\}} \prod_{i>j} (z_{n_i} - z_{n_j})^2 \prod_{i=1}^{M} z_{n_i}^J S_{n_i}^- |FM\rangle, \qquad (1.11)$$

where $z_n = \exp(i2\pi n/N)$, n_i is the down spin location, S^- the spin lowering operator, and $|FM\rangle$ is the ferromagnetic reference state. Here, N and M are the total number of spins and down spins, respectively, and J is the integer current of the down spin around the ring. The multicomponent generalization of the Haldane-Shastry model is obtained in [29, 31, 67].

The wavefunction (1.11) is that of the fully Gutzwiller projected free electrons and corresponds to a 1D version of the Kalmeyer-Laughlin wavefunction [60]. In the Dyson-Mehta-Gaudin theory of random matrices [17, 84], the wavefunction of the type given by (1.11) with n_i taken as a continuous variable is studied, and the corresponding correlation functions are known. The discreteness of the particle location does not affect some correlation functions; therefore, the results of Dyson, Mehta, and Gaudin are directly applicable to this model with some limitations.

The full eigenspectrum of both CSM and HSM (with $\lambda = 2$) is given by the following Bethe-ansatz-like equation

$$Nk_j = 2\pi I_j + \pi(\lambda - 1)\sum_{l=1}^{M} \text{sgn}(k_j - k_l), \tag{1.12}$$

where the pseudo-momenta k_j's and the integer or half-odd integer quantum numbers I_j are introduced for each down spin (particles) with $|I_j| \leq (N - M - 1)/2$ (no restrictions) for HSM (CSM). The total energy and (crystal-) momentum are given by

$$E = \sum_j k_j^2, \tag{1.13}$$

$$P = \sum_j k_j. \tag{1.14}$$

For HSM all the pseudo-momenta are translated by π, $k_j \to k_j + \pi$, such that the resulting momenta are in the range $(0, 2\pi)$. The interaction coupling constant λ is equal to 2 for the HSM. Note also that there is a striking equivalence between the energy spectra of the HSM and the corresponding CSM (i.e. $\lambda = 2$ case). A remarkable quantum symmetry called Yangian [16, 8] is responsible for the large degeneracy or the supermultiplets in the HSM [42] and, thus, for the equivalence to the CSM.

Haldane, first, verified numerically that Eq. (1.12) gives the complete energy spectra of the $SU(2)$ HSM and constructed a remarkable rule for obtaining the degeneracy for each energy level. The empirical rule allowed him to calculate the exact thermodynamics of this model [45]. The $SU(n)$ generalization of this rule is discussed in [42, 32], but it requires more detailed analysis. A different and complete method based on the idea of "squeezed strings" has been found in [32] and is explained later in this book. It has been found that the complete eigen-spectra of the $SU(n)$ HSM can be represented using the so called strings which are all "squeezed" onto the real axis, and that the string interaction responsible for their deformations in the nearest neighbor exchange (NNE) models are completely absent.

Perhaps the single most impressive feature of this class of model is that various dynamical correlation functions can be calculated exactly owing to fact that the elementary excitations in this class form ideal gas obeying fractional statistics. Details of this important feature are discussed in Chapter 7.

1.2.2 Hyperbolic and elliptic generalization

There is a further generalization of the inverse-square exchange (ISE) to the hyperbolic exchange. Inozemtsev has already suggested that the NNE and ISE Heisenberg models are two special limits of the hyperbolic model [55] whose hamiltonian is given by

$$H = J \sum_{i<j} \frac{P_{ij}}{\sinh^2(\gamma(i-j)/2)}. \tag{1.15}$$

This hamiltonian may be obtained by an analytic continuation of Eq. (1.10) (i.e. $N \to 2i/\gamma$) and can be shown to interpolate continuously between the NNE and ISE models. By taking the limit $\gamma \to \infty$ with $J \exp(-\gamma) = $ constant, the NNE Heisenberg model is obtained. On the other hand, the limit $\gamma \to 0$ with $J/\gamma^2 = $ constant gives the ISE hamiltonian which gives the Haldane-Shastry model if the periodic boundary condition is imposed. When the periodic boundary condition is imposed on the hamiltonian (1.15) the interaction becomes an elliptic function known as Weierstrass function.

All of the models discussed in this section and the previous section are classified as "integrable" models. When a model is "integrable" there are as many mutually commuting hamiltonians as the number of degrees of freedom in the model. For example, first few mutually commuting hamiltonians found for the Haldane-Shastry model in addition to the hamiltonian 1.10 are given by [42]

$$H_3 = \sum_{(ijk)}' \left(\frac{z_i z_j z_k}{z_{ij} z_{jk} z_{ki}} \right) P_{ijk}, \tag{1.16}$$

$$H_4 = \sum_{(ijkl)}' \left(\frac{z_i z_j z_k z_l}{z_{ij} z_{jk} z_{kl} z_{li}} \right) P_{ijkl} - 2\sum_{(ij)}' \left(\frac{z_i z_j}{z_{ij} z_{ji}} \right)^2 P_{ij}, \tag{1.17}$$

where P is the cyclic permutation operator. The primed sum omits equal values of the summation variables, and $z_{ij} \equiv z_i - z_j$, $z_j = \exp(i2\pi j/N)$. The forms of H_3 and H_4 for the HSM are similar to those for the NNE model given in Chapter 2. A systematic way of generating this series is given in [98]. A more formal way of saying the models are integrable is that they admit the Yang-Baxter equations. Recently, the Yang-Baxter structure of the ISE model is confirmed [42] using a formalism based on the Yangian algebra which is a formalized algebraic structure of the quantum Yang-Baxter equation. For the case of CSM and Haldane-Shastry model the explicit YBE has been found in [6]. These formal structures are not yet known for the elliptic case.

All of the models also share a common structure called the "strings" which correspond to simple patterns in the parameters called rapidities plotted in the complex plane. (Model systems in continuum space with no internal spin degrees of freedom allow only the simplest possible string structure, namely, the real rapidities or

1-strings.) The string structures for the NNE and ISE models are examined in chapters 2 and 5, and it is conjectured that the string structure for the elliptic models are somewhere in between the two extreme cases. Hence, it is tempting to conjecture that the string picture is universal and present in all of the models known to be integrable and, furthermore, is characteristic structure of systems supporting quantum solitonic excitations. It is also tempting to speculate that even in non-integrable 1D liquid phase the string structure is present with some characteristic life time depending on the strength of the integrability breaking.

1.3   Linearized models

In the previous two sections several explicit 1D models are introduced and briefly discussed. One of the advantages in using explicit models for studying 1D physics would be the relative ease of getting detailed, concrete and precise structures. This aspect is also a disadvantage since while some properties are quite robust some properties of the model system are less general and, therefore, less relevant to the experimentally attainable systems unless, of course, the model can be exactly reproduced.[5] Thus, one has to be very careful in making general statements based on the results found for some explicit model. One good example that illustrates this point is given by the integrable integer-spin chain whose low-energy spectrum is gap-less. Haldane, however, using more general method found that the integer spin chains have gapful low-energy spectra. It turns out that the integrable spin chain corresponds to some very special case. The solutions for the integrable integer chains are not wrong, but they are just less general and experimentally less relevant at present.

In this section several linearized 1D models *generally* valid in the zero temperature limit are discussed. These models when combined with the idea of renormalization group become very powerful in probing the zero-temperature behaviors of generic 1D liquid. I will start with a general many-body hamiltonian for 1D spinless fermions interacting with a pairwise potential given as follow

$$H = -\frac{\hbar^2}{2m} \int dx \psi^\dagger(x) \frac{\partial^2}{\partial^2 x} \psi(x) + \int dx dx' \psi^\dagger(x) \psi(x) V(x - x') \psi^\dagger(x') \psi(x'). \quad (1.18)$$

It is convenient to Fourier transform the above hamiltonian with the following transformation rules

$$\psi(x) = \frac{1}{\sqrt{L}} \sum_k e^{ikx} a_k, \quad (1.19)$$

$$\psi(x) = \frac{1}{\sqrt{L}} \sum_k e^{-ikx} a_k^\dagger, \quad (1.20)$$

$$V(x - x') = \frac{1}{L} \sum_q e^{i(x-x')q} V_q, \quad (1.21)$$

---

[5] As previously discussed the meaning of "unphysical" is really a function of the level of technology and, therefore, of time

where the operators a_k and $a_k^\dagger$ obey the usual anti-commutation relation $[a_k, a_{k'}^\dagger] = \delta_{kk'}$, and L is the size of the system with the usual periodic boundary condition. Note that since the potential is an even function, i.e. $V(x - x') = V(x' - x)$, its Fourier component is also an even function $V_q = V_{-q}$. The hamiltonian can be written in terms of the newly defined operators as

$$H = \sum_k e(k)a_k^\dagger a_k + \frac{1}{L}\sum_k V_k \rho_{-k}\rho_k, \qquad (1.22)$$

where the dispersion relations and the density operators ρ_k are given by

$$e(k) = \frac{\hbar^2 k^2}{2m}, \qquad (1.23)$$

$$\rho_q = \sum_k a_{k+q/2}^\dagger a_{k-q/2}. \qquad (1.24)$$

The general hamiltonian (1.22) cannot be solved exactly and, therefore, some approximations are needed. One such approximation is the linearization of the dispersion relation $e(k)$ near the Fermi points. As illustrated in Fig. 1.4 this approximation is valid in the limit of zero temperature since the particle-hole excitations are restricted to regions near the left and right Fermi points. The low-energy picture depicted here is clearly valid for a free system, but it is not clear whether the Fermi sea can survive once the interaction is turned on. It turns out that some relics of the Fermi sea survive and form a new type universal zero-temperature quantum fluid called "Tomonaga-Luttinger liquid" [37]. (The fluid is also called just Luttinger liquid.)

In the rest of this section I will illustrate the models explored first by Tomonaga and then by Luttinger. Tomonaga [103] first assumed that the dispersion is given by $e(k) = \hbar v_F |k|$ and then introduced the following density operators

$$\rho_k^+ = \sum_{q>0} a_{q-k/2}^\dagger a_{q+k/2}, \qquad (1.25)$$

$$\rho_k^- = \sum_{q<0} a_{q-k/2}^\dagger a_{q+k/2}. \qquad (1.26)$$

One can easily show that the commutation relations for the density operators evaluated over the ground state of the free system are given by

$$[\rho_k^+, \rho_{-k'}^+] = \frac{Lk}{2\pi}\delta_{kk'}, \qquad (1.27)$$

$$[\rho_k^-, \rho_{-k'}^-] = -\frac{Lk}{2\pi}\delta_{kk'}, \qquad (1.28)$$

$$[\rho_k^+, \rho_{-k'}^-] = 0, \qquad (1.29)$$

where $k > 0$. These commutation relations are valid so long as the particle distribution function do not change significantly as a result of the interaction.[6] If one defines

---

[6]The interaction in fact changes the particle distribution function qualitatively near the Fermi points but the overall change can be quite small; therefore, the commutation relations for the collective operators are not expected to change by very much.

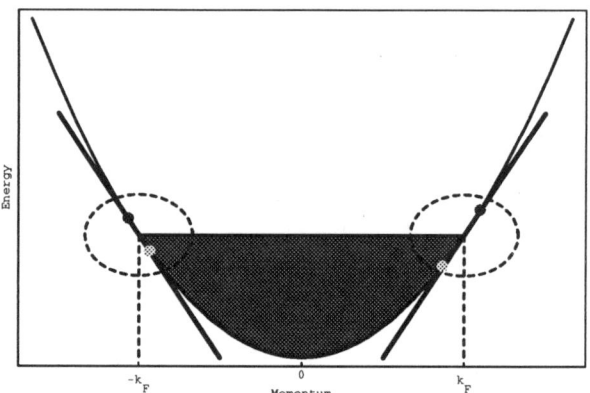

Fig. 1.4. Low-energy excitations in one-dimensional free Fermion gas where the shaded region indicates the filled Fermi sea.

the following new set of operators,

$$b_k = \sqrt{\frac{2\pi}{Lk}} \rho_k^+, \tag{1.30}$$

$$b_k^\dagger = \sqrt{\frac{2\pi}{Lk}} \rho_{-k}^+, \tag{1.31}$$

$$b_{-k}^\dagger = \sqrt{\frac{2\pi}{Lk}} \rho_k^-, \tag{1.32}$$

$$b_{-k} = \sqrt{\frac{2\pi}{Lk}} \rho_{-k}^-, \tag{1.33}$$

the commutation relations among the various density operators can be satisfied with a single bosonic one $[b_k, b_{k'}^\dagger] = \delta_{kk'}$. The new operators also satisfy the following commutations relations with the kinetic energy part of the hamiltonian, H_0,

$$[b_k, H_0] = \omega_k b_k, \tag{1.34}$$
$$[b_k^\dagger, H_0] = -\omega_k b_k^\dagger, \tag{1.35}$$

where $\omega = \hbar v_F |k|$. Therefore, at the level of commutation relations H_0 is just a system of simple non-interacting harmonic oscillators represented by the raising and lowering operators $b_k^\dagger$ and b_k, and the full hamiltonian is given by

$$H = \sum_k \omega_k b_k^\dagger b_k + \tilde{V}_k (b_k + b_{-k}^\dagger)(b_k^\dagger + b_{-k}), \tag{1.36}$$

where $\tilde{V}_k = |k|V_k/2\pi$. One can diagonalize the above hamiltonian with the following Bogoliubov transformation

$$b_k = (\cosh\gamma)\alpha_k + (\sinh\gamma)\alpha_{-k}^\dagger, \tag{1.37}$$

$$b_k^\dagger = (\cosh\gamma)\alpha_k^\dagger + (\sinh\gamma)\alpha_{-k}, \tag{1.38}$$

where γ is a free parameter, and α and $\alpha^\dagger$ are also bosonic operators. If the free parameter γ is set by

$$e^{4\gamma} = \frac{\omega_k}{\omega_k + 4\tilde{V}_k}, \tag{1.39}$$

the interacting system can again be represented by the simple harmonic oscillators, $H = \sum_k E_k \alpha_k^\dagger \alpha_k$, where the dispersion is modified to $E_k = \sqrt{\omega_k^2 + 4\omega_k \tilde{V}_k}$.

In order to make the various commutation relations in the Tomonaga model exact Luttinger introduced two sets of operators for the left- and right-movers near the Fermi points such that they obey the following anti-commutation relation

$$\{a_{k,\alpha}, a_{k',\alpha'}^\dagger\} = \delta_{\alpha\alpha'}\delta_{kk'}, \tag{1.40}$$

where α is the handedness index and can be either L or R. The density operators also acquire the handedness as

$$\rho_{k,\alpha} = \sum_p a_{p+k,\alpha}^\dagger a_{p,\alpha}, \tag{1.41}$$

where the sum is over all p. One can follow the same line of reasoning as shown above to construct and solve the Luttinger model. Unlike the Tomonaga model the Luttinger model is exactly solvable with no further approximation.

Haldane discovered that the Tomonaga-Luttinger model is robust against residual anharmonic couplings at low energies and, therefore, coined the term "Luttinger liquid" in analogy with the Fermi liquid [37].

Chapter 2

Heisenberg Spin Chain

As mentioned in Chapter 1 the Bethe-ansatz method has been used to exactly solve numerous one-dimensional quantum many-body systems. The isotropic spin-half Heisenberg chain is the first 1D many-body system that was solved by the Bethe-ansatz method by Bethe himself in 1931 [9]. Various people, thereafter, used the Bethe-ansatz to study the anti-ferromagnetic ground state [54], anisotropic generalization [87], excitations from the anti-ferromagnetic ground state [13], etc. The Bethe-ansatz has also been applied to the two-dimensional (2D) statistical mechanics problem such as the six-vertex model [76].

There are two most noteworthy achievements in the development of the Bethe-ansatz method since 1931. First is by C. N. Yang [109] who solved the one-dimensional particle system with the Dirac delta-function interaction (also solved independently by M. Gaudin [25]) and introduced the notion of factorizability of the N-body S-matrix into 2-body S-matrices. This work also provided the necessary mathematics for solution of the Hubbard model which is a lattice version of the Yang-Gaudin system [77]. Second is by R. Baxter who provided the transfer matrix formalism for the 2D eight-vertex model which can be mapped to the most general anisotropic XYZ spin-half Heisenberg chain [4]. These two works provided mathematical foundation for the Bethe-ansatz [97] and further had influenced other branches of mathematics including the quantum group [16, 8], the knot theory [62], etc.

In this chapter I use the $S = 1/2$ isotropic Heisenberg spin chain to describe the Bethe-ansatz method and use a simple and intuitive picture called "string" to represent the eigen-states of the nearest neighbor exchange (NNE) models. The "string" picture is exact only in the thermodynamic limit, but one can numerically show that even for a finite-size systems the strings provide useful descriptive language. This string picture becomes exact for the inverse-square exchange (ISE) model [32].

14

2.1 Elementary Features of the Heisenberg Spin Chain

In this section elementary language used to describe the spin one-half Heisenberg model is introduced. The Heisenberg hamiltonian is given by

$$H = 2J \sum_{\langle i,j \rangle} (\vec{S}_i \cdot \vec{S}_j - \frac{1}{4}), \tag{2.1}$$

where $\vec{S}_j$ is a local spin variable at jth site. $J > 0(< 0)$ corresponds to the anti-ferromagnetic (ferromagnetic) case. The sum is over the distinct nearest neighbors, and a constant term is added for convenience. For the spin-half case the spins are represented by the following Pauli matrices

$$S^x = \frac{1}{2}\sigma^1, \quad S^y = \frac{1}{2}\sigma^2, \quad S^z = \frac{1}{2}\sigma^3, \tag{2.2}$$

where

$$\sigma^1 = \begin{pmatrix} 0 & 1 \\ 1 & 0 \end{pmatrix}, \quad \sigma^2 = \begin{pmatrix} 0 & -i \\ i & 0 \end{pmatrix}, \quad \sigma^3 = \begin{pmatrix} 1 & 0 \\ 0 & -1 \end{pmatrix}. \tag{2.3}$$

The spin raising and lowering operators can be constructed as

$$S^+ \equiv S^x + iS^y = \begin{pmatrix} 0 & 1 \\ 0 & 0 \end{pmatrix}, \tag{2.4}$$

$$S^- \equiv S^x - iS^y = \begin{pmatrix} 0 & 0 \\ 1 & 0 \end{pmatrix}. \tag{2.5}$$

The eigen-states of the local spin operator S_i^z are given by

$$|\uparrow\rangle \equiv \begin{pmatrix} 1 \\ 0 \end{pmatrix}, \quad |\downarrow\rangle \equiv \begin{pmatrix} 0 \\ 1 \end{pmatrix}. \tag{2.6}$$

The many-body eigen-states of the total spin operator $S^z = \sum_j S_j^z$ are given by the following tensor product of the local states as

$$|v_1\rangle \otimes |v_2\rangle \otimes \cdots \otimes |\sigma_N\rangle \equiv |\sigma_1\sigma_2\ldots\sigma_N\rangle. \tag{2.7}$$

For example, when $N = 2$ there are total of four eigen-states of the total spin S^z: $|\uparrow\uparrow\rangle, |\uparrow\downarrow\rangle, |\downarrow\uparrow\rangle$ and $|\downarrow\downarrow\rangle$, and, in general, size of the state space (or Hilbert space) is 2^N.

The dot product of the spin operators can be re-expressed in terms of $S^\pm$ as

$$\begin{aligned} \vec{S}_i \cdot \vec{S}_j &= S_i^x S_j^x + S_i^y S_j^y + S_i^z S_j^z \\ &= \frac{1}{2}(S_i^+ S_j^- + S_i^- S_j^+) + S_i^z S_j^y. \end{aligned} \tag{2.8}$$

If I define P_{ij} as the spin exchange operator, then I have the following identity

$$\vec{S}_i \cdot \vec{S}_j = \frac{1}{2}(P_{ij} - \frac{1}{2}). \tag{2.9}$$

The original hamiltonian, therefore, is also given by

$$H = J\sum_{\langle i,j \rangle}(P_{ij} - 1). \tag{2.10}$$

For the case of two spins the four energy eigen-states are given by a singlet state, $|\uparrow\downarrow\rangle - |\downarrow\uparrow\rangle$, and three triplet states: $|\uparrow\uparrow\rangle$, $|\downarrow\downarrow\rangle$ and $|\uparrow\downarrow\rangle + |\downarrow\uparrow\rangle$.

The spin system above can also be described by lattice boson picture. One can make the following identifications:

$$|\downarrow\rangle \equiv |1\rangle, \quad |\uparrow\rangle \equiv |0\rangle,$$
$$S_i^+ = b_i, \quad S_i^- = b_i^\dagger, \quad S_i^z = \frac{1}{2} - b_i^\dagger b_i, \tag{2.11}$$

where $|1\rangle(|0\rangle)$ is a particle (hole) state and $b_i(b_i^\dagger)$ a particle destruction (creation) operator. The particle operators have a peculiar feature. They anti-commute on same site (i.e. $\{b_i, b_i^\dagger\} = b_i b_i^\dagger + b_i^\dagger b_i = 1$ and $(b_i)^2 = (b_i^\dagger)^2 = 0$), but commute on different sites. Thus, they are called "hard-core" bosons. The hamiltonian in terms of these hard-core boson operators is given by

$$H = \mathcal{P}\left(J\sum_{\langle i,j \rangle}(b_i^\dagger b_j + b_i b_j^\dagger) + 2J\sum_{\langle i,j \rangle}n_i n_j - 2JM\right)\mathcal{P}, \tag{2.12}$$

where $n_i = b_i^\dagger b_i$ and $\mathcal{P}$ a projection operator that projects out all the states with double and higher occupancy for any single site.

2.2 Bethe-Ansatz Solutions

In this section two methods for solving the Heisenberg spin model are presented. First method is based on the direct substitution of the Bethe's ansatz to the Schrödinger equation and the second on the $SU(2)$ highest weight condition.

The general eigen-state of the hamiltonian (2.10) can be represented by the eigen-states of $S^z = \sum_j S_j^z$ as

$$|\Psi\rangle = \sum_{1 \le n_i < \cdots < n_M \le N} \phi(n_1, \ldots, n_M)\prod_{j=1}^{M} S_{n_j}^- |FM\rangle, \tag{2.13}$$

where $S_{n_j}^-$ is the spin lowering operator at site n_j, and $|FM\rangle$ is the ferromagnetic reference state, i.e. $|\uparrow\uparrow \ldots \uparrow\rangle$. M is the total number of down spins and n_i the location of the down spin at ith site. The famous Bethe's ansatz that diagonalizes the spin hamiltonian is given by

$$\phi(n_1,\ldots,n_M) = \sum_P A_P \exp\left(i\sum_{j=1}^{M} k_{P_j}n_j\right), \tag{2.14}$$

where k_j's are the pseudo-momenta introduced for each of the M down spins, and P represents the permutation of $\{k_1,\ldots,k_M\}$. The total energy and crystal momentum are given in terms of the pseudo-momenta by

$$E = 2J\sum_{j=1}^{M}(\cos k_j - 1), \tag{2.15}$$

$$P = \sum_{j=1}^{M} k_j, \tag{2.16}$$

where $0 \leq k_j < 2\pi$.

2.2.1 Bethe-ansatz: Method I

The goal of this conventional Bethe-ansatz method is to find the coefficients A_P and all possible sets of allowed pseudo-momenta $\{k_j\}$. One way would be to substitute the ansatz (2.14) into the Schrödinger equation and solve the resulting difference equations as outlined below. Let me rewrite the hamiltonian in the following way

$$H = J(H_1 + H_2), \tag{2.17}$$

where

$$H_1 = \sum_{n=1}^{N} S_n^- S_{n+1}^+ + S_n^+ S_{n+1}^-, \tag{2.18}$$

$$H_2 = \sum_{n=1}^{N} 2(S_n^z S_{n+1}^z - \frac{1}{4}). \tag{2.19}$$

First, the action of H_1 on $|\Psi\rangle$,

$$\sum_{\{n\}} \phi(n_1,\ldots,n_M) \sum_{n=1}^{N} (S_n^- S_{n+1}^+ + S_n^+ S_{n+1}^-)|n_1,\ldots,n_M\rangle, \tag{2.20}$$

can be reorganized as

$$\sum_{\{n\}}\sum_{\{n\}'} \phi(\{n\}')|n_1,\ldots,n_M\rangle, \tag{2.21}$$

where the second sum is over all sets of down-spin positions $\{n\}'$ which are related to the set $\{n\}$ such that their corresponding spin configurations differ only by a nearest pair of up and down spins. For example, if $\{n\}$ is represented by the spin configuration $|\uparrow\uparrow\downarrow\uparrow\uparrow\downarrow\rangle$, then the sets $\{n\}'$ are given by $|\uparrow\downarrow\uparrow\uparrow\uparrow\downarrow\rangle$, $|\uparrow\uparrow\uparrow\downarrow\uparrow\downarrow\rangle$, and $|\uparrow\uparrow\downarrow\uparrow\downarrow\uparrow\rangle$. Thus, the total number of configurations belonging to the set $\{n\}'$ is equal to the total number of kinks (i.e. the up and down spin pairs) in the configuration $\{n\}$. The action of H_2 on $|\Psi\rangle$ simply gives a multiplicative numerical factor equal to the total number of kinks for any given spin configurations. Therefore, the Schrödinger equation becomes

$$\sum_{\{n\}'}(\phi(\{n\}') - \phi(\{n\})) = (E/J)\phi(\{n\}). \tag{2.22}$$

Before I discuss the general solution of Eq. (2.22), I want to examine the $M = 1$ and $M = 2$ cases. For $M = 1$ Eq. (2.22) reduces to

$$\phi(n+1) + \phi(n-1) - 2\phi(n) = (E/J)\phi(n), \tag{2.23}$$

whose solution is the following Block state

$$\phi(n) = Ae^{ikn}. \tag{2.24}$$

The energy $E = 2J(\cos(k) - 1)$ and, thus, the band-width is $|4J|$. By imposing the periodic boundary condition $\phi(n) = \phi(n + N)$ the allowed values of k are given by

$$k_j = \frac{2\pi}{N}\tilde{I}_j, \tag{2.25}$$

where the allowed quantum numbers I_j are $1, 2, \ldots, N$.

The $M = 2$ case is more complicated due to interactions between the down spins. I need to separately consider two cases, (i) $n_2 = n_1 + 1$ and (ii) $n_2 > n_1 + 1$, that provide the following two difference equations, respectively,

$$\begin{aligned}
\phi(n_1 - 1, n_1 + 1) &+ \phi(n_1, n_1 + 2) - 2\phi(n_1, n_1 + 1) \\
&= (E/J)\phi(n_1, n_1 + 1), \tag{2.26} \\
\phi(n_1 - 1, n_2) &+ \phi(n_1 + 1, n_2) + \phi(n_1, n_2 - 1) + \phi(n_1, n_2 + 1) \\
&- 4\phi(n_1, n_2) = (E/J)\phi(n_1, n_2). \tag{2.27}
\end{aligned}$$

The wavefunction ϕ has so far been defined only in region $1 \leq n_1 < n_2 \leq N$ where n_1 and n_2 are always distinct, but one can analytically extend the validity of ϕ to the region $1 \leq n_1 \leq n_2 \leq N$. Then, Eq. (2.26) is also given by Eq. (2.27) if the following auxiliary condition is imposed on ϕ

$$\phi(n_1 + 1, n_1 + 1) + \phi(n_1, n_1) - 2\phi(n_1, n_1 + 1) = 0. \tag{2.28}$$

This equation is a consistency (or boundary) condition for Eq. (2.27). If the following ansatz,

$$\phi(n_1, n_2) = Ae^{i(k_1 n_1 + k_2 n_2)} + Be^{i(k_2 n_1 + k_1 n_2)}, \tag{2.29}$$

is substituted into Eqs. (2.27) and (2.28), I get the eigen-energy and the ratio A/B as follow

$$E = 2J \sum_{j=1}^{2} (\cos(k_j) - 1), \qquad (2.30)$$

$$\frac{B}{A} \equiv e^{-i\psi_{12}} = -\frac{e^{i(k_1+k_2)} - 2e^{ik_2} + 1}{e^{i(k_1+k_2)} - 2e^{ik_1} + 1}. \qquad (2.31)$$

Note that $\psi_{12} = -\psi_{21}$. By imposing the periodic boundary condition $\phi(n_1, n_2) = \phi(n_2, n_1 + N)$, I get the following equations

$$e^{i(k_1 N - \psi_{12})} = 1, \qquad (2.32)$$

$$e^{i(k_2 N - \psi_{21})} = 1. \qquad (2.33)$$

The above equation can be re-written as

$$N k_j = 2\pi \tilde{I}_j + \psi_{jl}, \quad l \neq j, \qquad (2.34)$$

where $1 \leq \tilde{I}_j \leq N - 2$. By comparing the above equation with Eq. (2.25) one can see that the function ψ_{jl} corresponds to the scattering phase shift due to the spin-spin interaction.

It is straightforward to generalize the solution to M down spin case. For this purpose Eq. (2.22) is written more explicitly as follow

$$\sum_{j=1}^{M} \phi(n_1, \ldots, n_j + 1, \ldots, n_M) + \phi(n_1, \ldots, n_j - 1, \ldots, n_M) - 2\phi(n_1, \ldots, n_M)$$
$$= (E/J)\phi(n_1, \ldots, n_M). \qquad (2.35)$$

As in the $M = 2$ case there are unwanted terms, which stem from the states with nearest neighbor down spin pairs, in Eq. (2.35) which requires the following consistency condition,

$$\phi(\ldots, n_j + 1, n_j + 1, \ldots) + \phi(\ldots, n_j, n_j, \ldots) - 2\phi(\ldots, n_j, n_j + 1, \ldots) = 0. \qquad (2.36)$$

This consistency equation determines all the coefficients A_P in the ansatz given by Eq. (2.14) up to a multiplicative constant, and it is straightforward to show that they satisfy the following relation

$$\frac{A_{P(j,j+1)}}{A_P} \equiv e^{-i\psi_{j,j+1}} = -\frac{e^{i(k_{P_j}+k_{P_{j+1}})} - 2e^{ik_{P_{j+1}}} + 1}{e^{i(k_{P_j}+k_{P_{j+1}})} - 2e^{ik_{P_j}} + 1}, \qquad (2.37)$$

where $P(j, j+1)$ is equal to the permutation P with k_{P_j} and $k_{P_{j+1}}$ exchanged. The relation (2.37) can be easily satisfied by the following ansatz

$$A_P = e^{\frac{i}{2} \sum_{i<j} \psi_{i,j}}. \qquad (2.38)$$

From the following periodic boundary condition

$$\phi(n_1, n_2, \ldots, n_M) = \phi(n_2, \ldots, n_M, n_1 + N), \tag{2.39}$$

I obtain

$$A_{P'} = A_P e^{i k_{P_M} N}, \tag{2.40}$$

which translates into

$$e^{i k_j N} = \prod_{l(\neq j)} e^{i \psi_{jl}}. \tag{2.41}$$

By taking the logarithm of Eq. (2.41) and appropriately assigning different branches with integer multiples of 2π an equation corresponding to Eqs. (2.25) and (2.34) is obtained

$$N k_j = 2\pi \tilde{I}_j + \sum_{l \neq j} \psi_{j,l}, \tag{2.42}$$

where $\tilde{I}_j$ are integer quantum numbers in the range $1 \leq \tilde{I}_j \leq N - M$.

One can obtain another version of Eq. (2.42) by introducing a set of rapidity variables $\{\lambda_j\}$ which are related to the pseudo-momenta by

$$e^{i k_j} = (\lambda_j + i/2)/(\lambda_j - i/2), \tag{2.43}$$

or, equivalently, by

$$\lambda_j = \frac{1}{2} \cot(k_j/2). \tag{2.44}$$

This map between the fundamental regions in complex plane for $\tilde{k} = \pi - k$ and λ is illustrated in Fig. 2.1. The reparameterized pseudo-momenta $\tilde{k}$ reside in the infinite strip between $-\pi$ and $+\pi$ where as the rapidities λ takes on the entire complex plane. A labeled arrow in one plane gets mapped to the arrow with same label in the other plane. And, in terms of $\{\lambda_j\}$ Eq. (2.42) becomes

$$N\theta(2\lambda_j) = 2\pi I_j + \sum_{l} \theta(\lambda_j - \lambda_l), \tag{2.45}$$

where $\theta(x) = 2\mathrm{Arctan}(x)$, $|I_j| \leq (N - M - 1)/2$. And, Eq. (2.45) is rendered more symmetric about $\lambda = 0$ by making use of the following relation

$$\mathrm{Arccot}(x) = -\mathrm{Arctan}(x) + \frac{\pi}{2}. \tag{2.46}$$

The total energy and crystal momentum in terms of these new variables are given by

$$E = -J \sum_{j=1}^{M} \frac{1}{\lambda_j^2 + (1/2)^2}, \tag{2.47}$$

$$P = -2 \sum_{j=1}^{M} \mathrm{Arctan}(2\lambda_j) + M\pi \quad (\mathrm{mod}\ 2\pi). \tag{2.48}$$

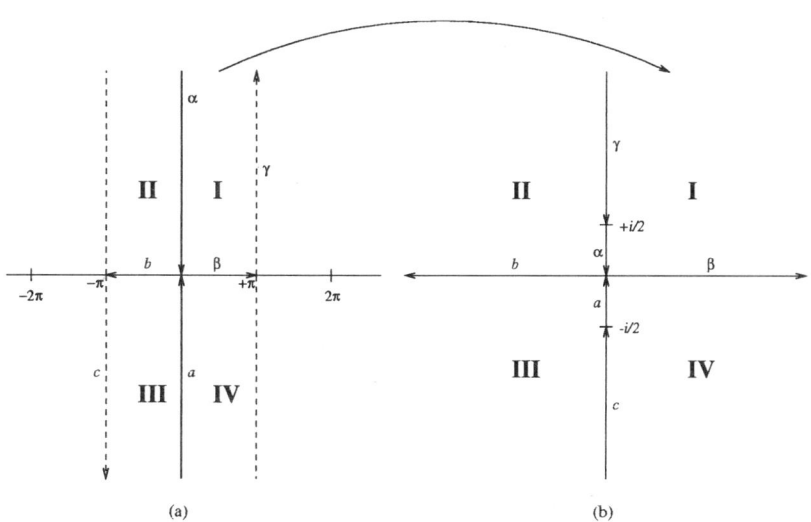

(a)   (b)

Fig. 2.1. Map $\frac{1}{2}\tan(\tilde{k}/2)$ between reparameterized pseudo-momentum $\tilde{k} = \pi - k$ and rapidity λ in complex plane. The labeled arrows in $\tilde{k}$-plane get mapped to arrows with same label in λ-plane. (a) The infinite strip bounded by the dashed lines represents the fundamental region for $\tilde{k} = \pi - k$. (b) Rapidities reside in the entire complex plane.

The solution of Eq. (2.45) for anti-ferromagnetic ground state in the infinite size limit (i.e. $N, M \to \infty$ with $M/N = 1/2$) is straightforward to obtain. When $M/N = 1/2$ all of the $N - M$ available quantum numbers are occupied so that $I_{j+1} - I_j = 1$ with $\{I_j\}$ in increasing order. It turns out that for $J > 0$ the minimum energy configuration corresponds to all real or 1-string rapidities symmetric about the zero axis (refer to section 2.5 for more discussion). Hence, $P = 0$ (π) for M even (odd) and the difference between the equations for λ_j and λ_{j+1} in the continuum limit becomes

$$a_1(\lambda) = 2\pi\sigma_1^0(\lambda) + \int_{-\lambda_0}^{\lambda_0} d\lambda' \sigma_1^0(\lambda') a_2(\lambda - \lambda'), \tag{2.49}$$

where $a_n(x) = n/(x^2 + (n/2)^2)$, and $\sigma_1^0(\lambda)$ is the ground state distribution function $(I_{j+1} - I_j)/N(\lambda_{j+1} - \lambda_j)$. Since all of the available rapidities are occupied when $M/N = 1/2$ for given N, one should set $\lambda_0 = \infty$ in which case Eq. (2.49) can be solved by Fourier transformation method. The distribution function, then, is given by

$$\tilde{\sigma}_1^0(\omega) = \int_{-\infty}^{\infty} d\lambda e^{-i\lambda\omega} \sigma_1^0(\lambda) = \frac{1}{2\cosh(\omega/2)}, \tag{2.50}$$

or

$$\sigma_1^0(\lambda) = \frac{1}{2\cosh(\pi\lambda)}. \tag{2.51}$$

One could easily check to see if $\int_{-\infty}^{\infty} d\lambda \sigma_1^0(\lambda) = 1/2$, and the ground state energy is given by

$$E_0 = -2JN \log 2. \tag{2.52}$$

Since the highest (lowest) energy eigenstate of the hamiltonian (2.1) or (2.10) for $J > 0$ $(J < 0)$ is the ferromagnetic state with zero energy E_0 is also the energy range of the model. It is interesting to note that this internal quantum energy difference as $N \to \infty$ is exactly same as the entropic energy as $T \to \infty$ with the following identifications: $\{N, 2J\} \equiv \{T, Nk_B\}$ where k_B is the Boltzmann constant. Therefore, one can consider the quantum ground (highest energy) state for $J > 0$ $(J < 0)$ NNE model as fully "thermalized" or fully quantum disordered with respect to the ferromagnetic state. In contrast the corresponding energy difference in the long-range interaction model given by the hamiltonian (1.10) in the large N limit is as follows:

$$\Delta E_0 = -2JN\frac{\pi^2}{12}. \tag{2.53}$$

There is about 20% difference between this and NNE case and it is apparently due to the residual quantum correlations in the ISE models. It is also interesting to note that this energy difference can be represented as the following series:

$$\begin{aligned} \Delta E_0 &= -2JN \sum_{k=1}^{\infty} \frac{(-1)^{k+1}}{k^s} \\ &= -2JN(1 - 2^{1-s})\zeta(s), \end{aligned} \tag{2.54}$$

where $s = 0, 1, 2$ corresponds to the Ising limit, NNE, and ISE, respectively. And, $\zeta(s)$ is the Riemann zeta function.

2.2.2 Bethe-ansatz: Method II

I show here another approach [79] which does not involve the hamiltonian (2.1) at all. I can self consistently determine all A_P and $\{k_j\}$ by imposing two conditions on the ansatz:

- (i) the periodic boundary condition (i.e. Eq. (2.39)),

- (ii) $S^+|\Psi\rangle = 0$, where S^+ is the total spin raising operator.

By imposing the periodic boundary condition on ϕ, I find the following relationship among A_P

$$A_{P'} = A_P \exp(ik_{P_M} N), \tag{2.55}$$

where $P_1' = P_M$, $P_2' = P_1$, $P_3' = P_2, \ldots, P_M' = P_{M-1}$. The condition (ii) restricts the eigen-states (2.13) to the $SU(2)$ highest weight states (i.e. $S = S^z$). The action of S^+ on $|\Psi\rangle$ is given by

$$\sum_{\{n\}} \sum_{i=1}^{M} \left(\sum_{j=n_{i-1}+1}^{n_i-1} \phi(n_1 \ldots, n_{i-1}, j, n_i, \ldots, n_{M-1}) \right) |n_1, \ldots, n_{M-1}\rangle, \tag{2.56}$$

where $n_0 \equiv 0$ and $n_M \equiv N + 1$. The third sum over j can be explicitly performed, and the $i = \nu - 1$ and $i = \nu$ terms are given by

$$\sum_P A_P \left(\frac{e^{ik_{P_{\nu-1}}(n_{\nu-2}+1)} - e^{ik_{P_{\nu-1}} n_{\nu-1}}}{1 - e^{ik_{P_{\nu-1}}}} \right) Q_{\nu-1}, \tag{2.57}$$

$$\sum_P A_P \left(\frac{e^{ik_{P_\nu}(n_{\nu-1}+1)} - e^{ik_{P_\nu} n_\nu}}{1 - e^{ik_{P_\nu}}} \right) Q_\nu, \tag{2.58}$$

where

$$Q_\nu = \left(\prod_{l=1}^{\nu-1} e^{ik_{P_l} n_l} \right) \left(\prod_{m=\nu}^{M-1} e^{ik_{P_{m+1}} n_m} \right). \tag{2.59}$$

The sums in Eqs. (2.57) and (2.58) can be divided into two parts and the second part in the former and the first part in the latter equation can be summed as follow

$$\sum_P A_P \left(\frac{-1}{1 - e^{ik_{P_{\nu-1}}}} + \frac{e^{ik_{P_\nu}}}{1 - e^{ik_{P_\nu}}} \right) e^{i(k_{P_{\nu-1}}+k_{P_\nu})n_{\nu-1}} \tilde{Q}_\nu, \tag{2.60}$$

where $\tilde{Q}_\nu = Q_\nu / \exp(ik_{P_{\nu-1}} n_{\nu-1})$. The sum above can be reorganized by adding the permutation $P' = P(\nu - 1, \nu)$ dependent terms to the P dependent terms to give

$$\sum_P \left[A_P \left(\frac{-1}{1 - e^{ik_{P_{\nu-1}}}} + \frac{e^{ik_{P_\nu}}}{1 - e^{ik_{P_\nu}}} \right) + A_{P'} \left(\frac{-1}{1 - e^{ik_{P_\nu}}} + \frac{e^{ik_{P_{\nu-1}}}}{1 - e^{ik_{P_{\nu-1}}}} \right) \right]$$
$$\times e^{i(k_{P_{\nu-1}}+k_{P_\nu})n_{\nu-1}} \tilde{Q}_\nu. \tag{2.61}$$

In order for this sum to vanish the terms in the square bracket must vanish, and in that case I get the following relationship between A_P and $A_{P'}$

$$\frac{A_{P'}}{A_P} = -\frac{1 - 2e^{ik_{P_\nu}} + e^{i(k_{P_{\nu-1}} + k_{P_\nu})}}{1 - 2e^{ik_{P_{\nu-1}}} + e^{i(k_{P_{\nu-1}} + k_{P_\nu})}}, \tag{2.62}$$

which is identical to the relationship found before in Eq. (2.63). The periodic boundary condition is necessary to cancel extra unwanted terms in Eq. (2.56) and this is left as an exercise for the readers.

Once a coefficient, say A_I, is fixed, all the other coefficients are obtained by successively applying the permutation operators in conjunction with Eqs. (2.55) and (2.62). I combine the two relations in Eqs. (2.55) and (2.62) and introduce a new set of parameters called rapidities given by $\exp(ik_j) = (\lambda_j + i/2)/(\lambda_j - i/2)$ as before and obtain the following Bethe-ansatz equations (BAE),

$$\left(\frac{\lambda_j + \frac{i}{2}}{\lambda_j - \frac{i}{2}}\right)^N = -\prod_{l=1}^{M} \frac{\lambda_j - \lambda_l + i}{\lambda_j - \lambda_l - i}. \tag{2.63}$$

The self-consistent solution of Eq. (2.63) would give the pseudo-momenta $\{k_j\}$ via the rapidity-momentum map Eq. (2.44).

Two puzzles might cross one's mind at this point. (i) Why does the ansatz given by Eq. (2.14) diagonalize the hamiltonian (2.1)—after all, the hamiltonian is not necessary in determining the BAE (2.63)? (ii) Since Eq. (2.63) can be written as a polynomial of order $N + M$ for λ_j for a given set of $\{\lambda_l\}$ with $l \neq j$, there should be extra N roots for λ_j not belonging to the set $\{\lambda_j\}$. What are the other N roots for λ_j? These questions will be answered in the following section.

2.3 Transfer Matrix Formalism

In this section we give only some essential elements of the formalism based on the transfer matrix method also known as the algebraic Bethe-ansatz. For more details the readers are referred to, for example, [4, 97].

The transfer matrix, which is essentially a N-body scattering matrix, is defined as follow

$$T(\lambda) = \text{Tr}(L_N(\lambda) \cdots L_1(\lambda)), \tag{2.64}$$

where Tr denotes the usual trace of a matrix, and L_n is a local operator having the following form

$$L_n(\lambda) = \begin{pmatrix} \lambda + iS_n^z & iS_n^- \\ iS_n^+ & \lambda - iS_n^z \end{pmatrix}. \tag{2.65}$$

Here, $S^\pm$ are the usual spin raising and lowering operators, and λ is a complex number called the spectral parameter which turns out to be equivalent to the rapidity variables defined in the preceding section.

One can show that the eigenvalues of the transfer matrix $T(\lambda)$ are given by

$$\Lambda(\lambda; \{\lambda_j\}) = (\lambda + \frac{i}{2})^N \frac{Q(\lambda - i)}{Q(\lambda)} + (\lambda - \frac{i}{2})^N \frac{Q(\lambda + i)}{Q(\lambda)}, \qquad (2.66)$$

where

$$Q(\lambda) = \prod_{j=1}^{M} (\lambda - \lambda_j), \qquad (2.67)$$

provided the unequal parameters $\{\lambda_j\}$ satisfy Eq. (2.63). Hence, the set $\{\lambda\}$ is just the spin rapidity introduced in the previous section, and the BAE (2.63) is a consistency condition required for Λ being the eigenvalues of the transfer matrix. Solution of the BAE corresponds to the zeroes of Q, and Eq. (2.63) insures that the residues of the simple poles of Λ vanish. Λ is a polynomial of order N, and the zeroes of Λ denoted by $\{\bar{\lambda}_j\}$ are the "missing" roots mentioned in the previous section.

The transfer matrix, furthermore, has the following commutation relation

$$[T(\lambda), T(\mu)] = 0, \qquad (2.68)$$

for any λ and μ as long as there exists a non-singular 4×4 number matrix $R(\lambda)$ such that

$$R(\lambda - \mu)(L_n(\lambda) \otimes L_n(\mu)) = (L_n(\mu) \otimes L_n(\lambda))R(\lambda - \mu), \qquad (2.69)$$

where $\otimes$ is the tensor product of the two 2×2 matrices, i.e.

$$\begin{pmatrix} a & b \\ c & d \end{pmatrix} \otimes \begin{pmatrix} e & f \\ g & h \end{pmatrix} = \begin{pmatrix} a\begin{pmatrix} e & f \\ g & h \end{pmatrix} & b\begin{pmatrix} e & f \\ g & h \end{pmatrix} \\ c\begin{pmatrix} e & f \\ g & h \end{pmatrix} & d\begin{pmatrix} e & f \\ g & h \end{pmatrix} \end{pmatrix}. \qquad (2.70)$$

R. Baxter [4] found such matrix for the eight vertex model (which can be mapped to the Heisenberg XYZ chain). For our simple XXX (i.e. isotropic) chain the R matrix is given by

$$R(\lambda) = \begin{pmatrix} 1 & 0 & 0 & 0 \\ 0 & b(\lambda) & c(\lambda) & 0 \\ 0 & c(\lambda) & b(\lambda) & 0 \\ 0 & 0 & 0 & 1 \end{pmatrix}, \qquad (2.71)$$

where $b(\lambda) = i/(\lambda + i)$ and $c(\lambda) = \lambda/(\lambda + i)$. This feature of the transfer matrix elegantly expressed by Eq. (2.69) is what makes the entire formalism work!

The transfer matrix $T(\lambda)$ is the generating function for the infinite number of mutually commuting operators, and the hamiltonian (2.1) is just one of them. Since the eigenvalue of the transfer matrix $\Lambda(\lambda)$ can be written as $\prod(\lambda - \bar{\lambda}_j)$, the eigenvalues of $\log T(\lambda)$ and of the higher order derivatives evaluated at some λ_0 would be additive (i.e. $\sum_j f(\bar{\lambda}_j)$). Let us define the following operators

$$H_{(k+1)} = i^k \frac{d^k}{d\lambda^k} \log T(\lambda)\bigg|_{\lambda=\lambda_0}. \qquad (2.72)$$

A complete set of mutually commuting hamiltonian is given by the above sequence. For example, H_2 with $\lambda_0 = i/2$ corresponds to the Heisenberg hamiltonian (2.1). H_3 and H_4 are given by

$$H_3 = \sum_{i=1}^{N}(P^+_{i,i+1,i+2} - P^-_{i,i+1,i+2}), \tag{2.73}$$

$$H_4 = \sum_{i=1}^{N}(P^+_{i,i+1,i+2,i+3} + P^-_{i,i+1,i+2,i+3} - Q^+_{i,i+1,i+2,i+3} - Q^-_{i,i+1,i+2,i+3}$$
$$+2P_{i,i+1} - P_{i,i+2}), \tag{2.74}$$

where P^+ is the cyclic permutation operator and P^- its inverse. P is the usual exchange operator, and $Q^\pm$ are mutually inverse operators defined by the following actions: $Q^+(ijkl) = (jlik)$ and $Q^-(ijkl) = (kilj)$. One can easily check that H_2, H_3, and H_4 mutually commute. H_n, in general, has the n-body permutation operators and the residual r-body operators with $r \leq n$.

2.4 Strings in the Heisenberg Chain

Haldane in his unpublished work [41] obtained the eigenvalues of the transfer matrix by direct numerical diagonalization and numerically solved Eq. (2.66) for Q. The zeroes of T and Q for the Heisenberg chain with 12 sites are plotted in the complex plane as "o" and "x", respectively, in Fig. 2.2. The λ's ("x") lined up and separated along the imaginary axis approximately by $i = \sqrt{-1}$ are called a "string." The zeros of the transfer matrix ($\bar{\lambda}$) are usually located approximately $\pm i$ above and below the λ string along the imaginary axis. The ground state string configuration is shown in Fig. 2.2(a). The 2-string and 1-string are on top of each other in Fig. 2.2(b), and two pairs of $\bar{\lambda}$'s are lined up along with the strings. Unlike the 2- and 1-string, the 3- and 1-string do not "get along" with each other as shown by the distortions in Fig. 2.2(c). There are two λ's on the real axis symmetric about the origin, and it is difficult to see which one belongs to the 3-string. In Fig. 2.2(d) a similar situation occurs but we can unambiguously determine which of the two real λ's belong to the 3-string despite the distortion. In Fig. 2.2(d) a spectacular display of a 4-string can be observed. The distortion of the strings can occur not only along the real axis but also along the imaginary axis as shown in Fig. 2.2(f) where 1-, 2-, and 3-strings are all on top of each others. The 2-string is not distorted; however, the 3- and 1-string are deformed mostly along the imaginary axis near the origin. It is also difficult to properly assign the pairs of $\bar{\lambda}$'s to the corresponding strings.

I conclude that most of the deformations of strings are introduced by the "interaction" between the even- and even-string (E-E) or the odd- and odd-string (O-O). The even- and odd-string (E-O) do not seem to distort each other very much. There is a very intuitive explanation for this behavior. The λ's have to be all unequal numbers. (Whenever two λ's are equal, the corresponding wavefunction vanishes.) The

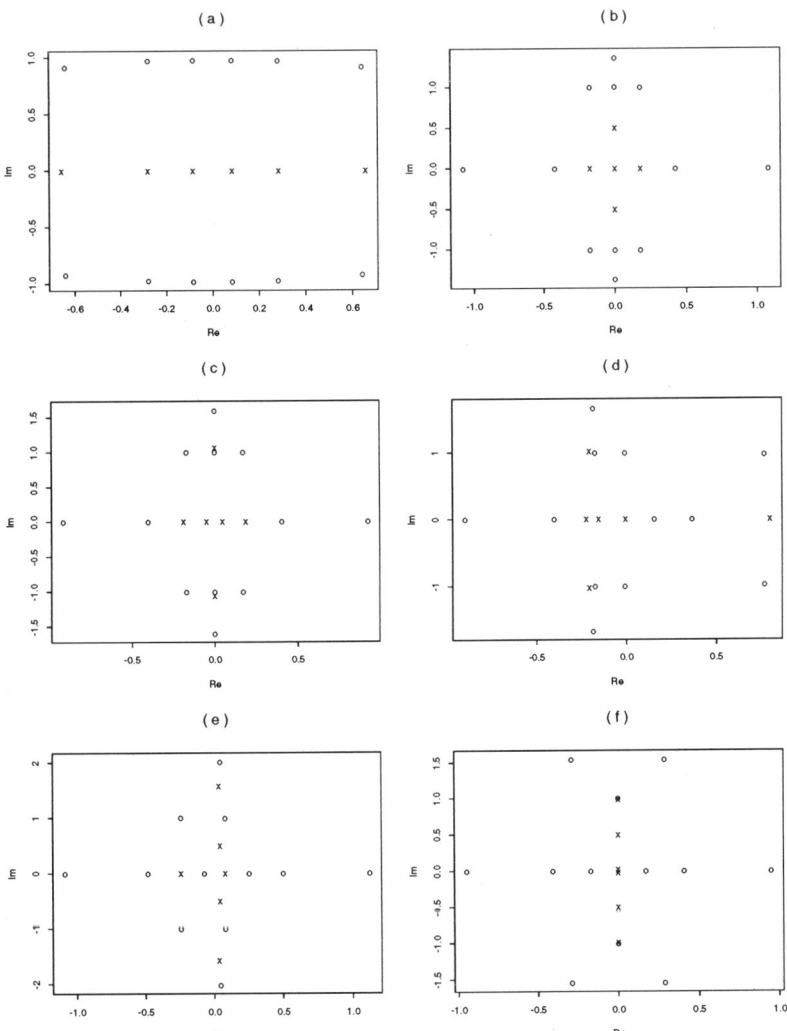

Fig. 2.2.   The λ("x") and $\bar{\lambda}$("o") are plotted in the complex plane. A set of "x" lined up and separated along the imaginary axis by i are called a string. There is a pair of "o" associated with each string. (a) The ground state; (b) 2- and 1-string on top of each other; (c),(d) Distortion introduced by 3- and 1-string "interaction"; (e) 4-string; (f) 1-, 2-, and 3-string line-up.

E-O strings can be on top of each other without any pair of the λ's being equal. On the other hand, the E-E or O-O strings cannot.

Despite the obvious distortions the string picture seems to be a characteristic feature of the eigen-spectra of the Heisenberg spin chain. It is generally believed that this picture becomes exact in the thermodynamic limit.

2.5 Thermodynamic Bethe-Ansatz Equations

Some numerical solution of the BAE (2.63) are shown in the previous section to be in the form of strings. In this section a systematic approach based on this string picture is developed for obtaining the thermodynamics of the Heisenberg spin chain.

The n-strings are defined as follow

$$\lambda_{\alpha,j}^n = \lambda_\alpha^n + \frac{i}{2}(n+1-2j), \tag{2.75}$$

where $j = 1, 2, \ldots, n$ and λ_α^n is a real parameter that is used to represent the n-string. The energy of the system in terms of the string variables is given by

$$E = -J \sum_{n=1}^{M} \sum_\alpha a_n(\lambda_\alpha^n), \tag{2.76}$$

where

$$a_n(\lambda) = \frac{n}{\lambda^2 + (n/2)^2}. \tag{2.77}$$

Hence, $-Ja_n(\lambda)$ corresponds to the energy function of n-string and since $-(Ja_n(\lambda))/n$ for $J > 0$ is always greater than that for shorter strings at the same value of λ the configurations with shorter strings usually have less energy.

In order to re-express the BAE in terms of the string variables it is convenient to multiply the BAE for each of the n $\lambda_{\alpha,j}^n$'s in a given string of length n. The left side of the resulting BAE is given by

$$\prod_{j=1}^{n} \frac{\lambda_{\alpha,j}^n + i/2}{\lambda_{\alpha,j}^n - i/2} = \prod_{j=1}^{n} \frac{\lambda_\alpha^n + (n+1-(2j-1))i/2}{\lambda_\alpha^n + (n+1-(2j+1))i/2}$$

$$= \frac{\lambda_\alpha^n + ni/2}{\lambda_\alpha^n - ni/2}. \tag{2.78}$$

The right-hand side the BAE is given by

$$\left[\prod_{j \neq l} \frac{\lambda_{\alpha,j}^n - \lambda_{\alpha,l}^n + i}{\lambda_{\alpha,j}^n - \lambda_{\alpha,l}^n - i}\right] \left[\prod_\beta \prod_{m=1}^{M} \prod_{j=1}^{n} \prod_{l=1}^{m} \frac{\lambda_{\alpha,j}^n - \lambda_{\beta,l}^m + i}{\lambda_{\alpha,j}^n - \lambda_{\beta,l}^m - i}\right], \tag{2.79}$$

where the first factor comes from the roots in the same string and the second from all the other strings. The first factor simplifies to

$$\prod_{j \neq l} \frac{(n+1-2j)i/2 - (n+1-2l)i/2 + i}{(n+1-2j)i/2 - (n+1-2l)i/2 - i} = \prod_{j \neq l} \frac{l-j+1}{l-j-1}$$

$$= \left[\prod_{j>l} \frac{l-j+1}{l-j-1} \right] \left[\prod_{j>l} \frac{l-j-1}{l-j+1} \right]$$

$$= 1. \tag{2.80}$$

The second factor needs more calculations and is given as follow

$$\prod_{\beta}^{M} \prod_{m=1} \left[\prod_{j=1}^{n} \prod_{l=1}^{m} \frac{\lambda_\alpha^n - \lambda_\beta^m + i((n-m)/2 + l - j + 1)/2}{\lambda_\alpha^n - \lambda_\beta^m + i((n-m)/2 + l - j - 1)/2} \right.$$

$$= \prod_{j=1}^{n} \frac{(\lambda_\alpha^n - \lambda_\beta^m + i((n+m)/2 - j))(\lambda_\alpha^n - \lambda_\beta^m - i((n-m)/2 - j))}{(\lambda_\alpha^n - \lambda_\beta^m - i((n+m)/2 - j))(\lambda_\alpha^n - \lambda_\beta^m + i((n-m)/2 - j))}$$

$$= \prod_{j=0}^{\min(n,m)} \left. \left(\frac{\lambda_\alpha^n - \lambda_\beta^m + i(|n-m|/2 + j)}{\lambda_\alpha^n - \lambda_\beta^m - i(|n-m|/2 + j)} \right)^{2-\delta_{j,0}-\delta_{j,\min(n,m)}} \right]. \tag{2.81}$$

By taking the logarithm of the BAE one gets the following form

$$N\theta(2\lambda_\alpha^n/n) = 2\pi J_\alpha^n + \sum_{\beta,m} \Theta_{nm}(\lambda_\alpha^n - \lambda_\beta^m), \tag{2.82}$$

where

$$\theta(x) = 2\text{Tan}^{-1}(x), \tag{2.83}$$

$$\Theta_{nm}(x) = (1 - \delta_{n,m})\theta(2x/|n-m|) + \theta(2x/(n+m))$$

$$+ 2 \sum_{j=1}^{\min(n,m)-1} \theta(2x/(|n-m|+2j)). \tag{2.84}$$

The quantum number is restricted to

$$|J_\alpha^n| \leq \frac{1}{2}(N - \sum_{m=1}^{M} c_{nm}\nu_m - 1), \tag{2.85}$$

where $c_{nm} = 2\min(n,m) - \delta_{nm}$ and ν_m is the total number of m-strings such that $\sum_{m=1}^{M} m\nu_m = M$. A set of solution $\{\lambda^n\}$ of Eq. (2.82) corresponding to a given set of quantum numbers $\{J_p^n\}$ is called the "particle" solution; and the solution corresponding to the complimentary set of unused quantum numbers $\{J_h^n\}$ (i.e. $\{J_p^n\} + \{J_h^n\} = \{J^n\}$ where $\{J^n\}$ is given by Eq. (2.85)) is called the "hole" solution.

The total number of states generated from the strings is equal to the total number of ways for assigning a distinct quantum number to each string in a given state. Since the total number of available quantum numbers for n-strings is $N - \sum_m c_{nm}\nu_m$, the total number of states for a given configuration of $\{\nu_n\}$ is

$$\prod_{n=1}^{M} \binom{N - \sum_m c_{nm}\nu_m}{\nu_n}. \tag{2.86}$$

One can show that the above sum is equal to [99]

$$\binom{N}{M} - \binom{N}{M-1},\tag{2.87}$$

where the binomial coefficient with negative entree is defined to be zero. Since the degeneracy of the $SU(2)$ highest weight states is given by $2S + 1 = 2S^z + 1 = N - 2M - 1$, the total number of states is given by

$$\sum_{M=0}^{[N/2]} (N - 2M + 1) \left\{ \binom{N}{M} - \binom{N}{M-1} \right\} = 2^N,\tag{2.88}$$

where $[N/2]$ is equal to $(N-1)/2$ $(N/2)$ if N is odd (even). Therefore, one can conclude that a complete set of $SU(2)$ highest weight states is given by the string picture.

In the limit N, $M \to \infty$ with M/N finite the equations can be written as integrals, and the number densities $\sigma_n(\lambda)$ and $\sigma_n^h(\lambda)$ for the "particle" and the "hole" solutions can be defined as follow. If the set $\{J_p^n\}$ is an ordered set the following relationship must hold in the thermodynamic limit

$$\frac{J_{p,\alpha+1}^n - J_{p,\alpha}^n}{N(\lambda_{\alpha+1}^n - \lambda_\alpha^n)} \to \sigma_n(\lambda) + \sigma_n^h(\lambda).\tag{2.89}$$

Thus, by taking the first derivative of the thermodynamic Bethe-ansatz equation I obtain the following expression

$$a_n(\lambda) = 2\pi\sigma_n^h(\lambda) + \sum_{m=1}^{\infty} \int_{-\infty}^{\infty} d\lambda' \sigma_m(\lambda') A_{nm}(\lambda - \lambda'),\tag{2.90}$$

where

$$a_j(x) = \begin{cases} 2\pi\delta(x) & \text{for } j = 0, \\ j/(x^2 + (j/2)^2) & \text{for } j \neq 0, \end{cases}\tag{2.91}$$

$$A_{nm}(x) = a_{|n-m|}(x) + a_{n+m}(x) + 2\sum_{j=1}^{\min(n,m)-1} a_{|n-m|+2j}(x).\tag{2.92}$$

Note that if the function $a_j(x)$ is analytically continued for the non-integer values of j it would converge to $2\pi\delta(x)$ as $j \to 0$ and the term $2\pi\sigma_n(\lambda)$ in Eq. (2.90) is absorbed into the definition of A_{nm}. These functions are identical to the energy functions given by Eq. (2.77).

In order to calculate the equilibrium distribution of the strings at finite temperature I need to minimize the free energy with respect to the distribution functions σ_n and σ_n^h. The free energy is defined as $F = E - TS - hS^z$ where h is the external magnetic field, T the temperature and S the entropy. The total internal energy E and the Zeeman energy, $-hS^z$, are given by

$$E/N = -J \sum_{n=1}^{\infty} \int_{-\infty}^{\infty} d\lambda \sigma_n(\lambda) a_n(\lambda), \qquad (2.93)$$

$$-hS^z/N = -\frac{h}{2} + h \sum_{n=1}^{\infty} \int_{-\infty}^{\infty} d\lambda n \sigma_n(\lambda). \qquad (2.94)$$

From Eq. (2.93) one sees that $a_n(\lambda)$ is the energy distribution function In order to find the entropy I first need to know the total number of configurations which can be calculated in the following way. Consider a box of size $\Delta\lambda$ in the rapidity space such that $\sigma_n(\lambda)\Delta\lambda$ and $\sigma_n^h(\lambda)\Delta\lambda$ (i.e. the number of "particles" and "holes" in the box) are large compared to unity and do not change much with respect to small variations in size of the box. Then, the total number of configurations is equal to the product over all such boxes and string length

$$\prod_{n=1}^{\infty} \prod_{\lambda} \frac{[(\sigma_n(\lambda) + \sigma_n^h(\lambda))\Delta\lambda]!}{(\sigma_n(\lambda)\Delta\lambda)!(\sigma_n^h(\lambda)\Delta\lambda)!}, \qquad (2.95)$$

where the product index λ is the box label. The entropy per site, therefore, is given by

$$S/N = \sum_{n=1}^{\infty} \int_{-\infty}^{\infty} d\lambda \left[\left(\sigma_n(\lambda) + \sigma_n^h(\lambda)\right) \log \left(\sigma_n(\lambda) + \sigma_n^h(\lambda)\right) \right.$$
$$\left. - \sigma_n(\lambda) \log \sigma_n(\lambda) - \sigma_n^h(\lambda) \log \sigma_n^h(\lambda) \right], \qquad (2.96)$$

where the Boltzmann constant is taken to be unity. By minimizing the free energy with respect to the variations in σ_n and σ_n^h and using Eq. (2.90) it is straightforward to obtain the following coupled integral equations

$$\log(1 + \eta_n(\lambda)) = \frac{nh}{T} - \frac{J}{T} a_n(\lambda) + \sum_{m=1}^{\infty} \int \frac{d\lambda'}{2\pi} A_{nm}(\lambda - \lambda') \log(1 + \eta_m^{-1}(\lambda')), \qquad (2.97)$$

where $\eta_n = \sigma_n^h/\sigma_n$. I can also re-express the above relations in a more compact form as

$$\log \eta_n(\lambda) - \hat{s} \log \left[(1 + \eta_{n-1}(\lambda))(1 + \eta_{n+1}(\lambda)) \right], \qquad (2.98)$$

where the operator $\hat{s}$ is defined as

$$\hat{s} f(\lambda) = \int_{-\infty}^{+\infty} d\lambda' \frac{f(\lambda')}{2 \cosh(\pi(\lambda - \lambda'))}. \qquad (2.99)$$

There are also the following two boundary conditions for the difference equations

$$\log(1 + \eta_0(\lambda)) = -\frac{2\pi J}{T} \delta(\lambda), \qquad (2.100)$$

$$\lim_{n \to +\infty} \frac{\log \eta_n}{n} = \frac{h}{T}. \qquad (2.101)$$

Note that the condition (2.100) is an analytical extension of Eq. (2.97) with $A_{0m} \equiv 0$. With some manipulation the free energy can be express as follow

$$F/N = E_0/N - T \int_{-\infty}^{\infty} d\lambda \sigma_1^0(\lambda) \log[1 + \eta_1(\lambda)], \qquad (2.102)$$

where $E_0 = -2JN \log 2$ (the ground state energy), and $\sigma_1^0(\lambda) = 1/2 \cosh(\pi\lambda)$ (the ground state distribution function). It is interesting to note that this finite temperature free energy is expressed only in terms of 1-string functions $\eta_1(\lambda)$ and $\sigma_1^0(\lambda)$.

In order to obtain the free energy η_1 needs to be calculated from the system of difference equations which cannot be solved in general but some asymptotic solutions can be obtained. Details of the asymptotic solutions will not be presented here; however, similar equations for the Hubbard model are solved in Chapter 3. The thermodynamic equations in this section get greatly simplified for the inverse-square exchange models as shall be shown in Chapter 5.

Chapter 3

The 1D Hubbard Model

3.1 Introduction

The Hubbard model [53] describing the strongly correlated electron systems are of great physical interest. The high-temperature superconductivity, for example, is believed to be described by the two-dimensional (2-D) repulsive Hubbard model near the half-filled band [1]. The model in more than one-dimension has not been solved and, therefore, one hopes to learn some features of the 2-D model by studying the exact solutions of the one-dimensional Hubbard model. In this chapter I concentrate on thermodynamic properties of the 1-D model using the string picture.

In 1967 C.N. Yang [109] and M. Gaudin [25] used the Bethe ansatz to solve one dimensional electron system with delta-function interaction. Lieb and Wu [77] generalized Yang's solution to a lattice case (i.e. the 1D Hubbard model) and obtained the exact ground state energy of the model. Several authors [92, 100, 14], thereafter, investigated other ground state properties. Takahashi [100] used Lieb and Wu's solution to obtain the finite-temperature Bethe-ansatz equations for the model and showed that the equations predict the correct thermodynamic behaviors in four different special cases—namely, (i) infinite and (ii) zero coupling limits, and (iii) infinite and (iv) zero temperature limits. In deriving the thermodynamic Bethe-ansatz equations (TBAE) the so called "string hypothesis" is used where the excited states are characterized by "strings."[1] Even though the hypothesis of this type were used by many authors [9, 63, 25], it has never been proved rigorously and the string has not received much attention for nearly two decades.

In this chapter I introduce the strings for the charge and spin complexes and develop a new type of highly convergent strong-coupling expansion series (λ-expansion). Some thermodynamic properties of the model are also investigated using the λ-expansion.

---

[1]Reminder: A set of complex rapidities that share a common real part and are separated along the imaginary axis by $\sqrt{-1}$ is called a string.

3.2 Bethe-Ansatz Equations

The hamiltonian of the model written in a symmetric form is given by

$$H = - \sum_i \sum_\sigma (c_{i\sigma}^\dagger c_{i+1\sigma} + c_{i\sigma}^\dagger c_{i+1\sigma}) + U \sum_i (n_{i\uparrow} - \frac{1}{2})(n_{i\downarrow} - \frac{1}{2}), \qquad (3.1)$$

where $c_{i\sigma}^\dagger, c_{i\sigma}$, and $n_{i\sigma}$ are the creation, annihilation, and number operators, respectively, for an electron at site i with spin σ ($\uparrow$ or $\downarrow$). The transfer integral, t, is set to unity and the on-site interaction energy is parameterized by U.

The hamiltonian (3.1) is invariant under the following unitary transformations: $c_{i\sigma} \to \sum_{\sigma'} U_{\sigma\sigma'} c_{i\sigma'}$ and $c_{i\sigma} \to e^{i\theta} c_{i\sigma}$, where U is an $SU(2)$ matrix. The Hubbard model, therefore, has a $SU(2) \times U(1)$ symmetry. Furthermore, when the number of sites is even, there is a "hidden" $SU(2)$ symmetry where the $U(1)$ charge symmetry is a subgroup of the hidden $SU(2)$. When the number of sites is odd, a more general twisted boundary condition is needed to get the extra $SU(2)$ symmetry. The generators of this hidden pseudo-spin symmetry is as follow [39]

$$Q^z = \frac{1}{2}(N_a - \sum_{i,\sigma} c_{i\sigma}^\dagger c_{i\sigma}), \qquad (3.2)$$

$$Q^+ = \sum_i (-1)^i c_{i\uparrow} c_{i\downarrow}, \qquad (3.3)$$

$$Q^- = (Q^+)^\dagger. \qquad (3.4)$$

(Here, I use different but essentially equivalent notations as those used in [111, 19].) Essentially, the extra symmetry appears between the doubly occupied and empty sites which correspond to the up and down states for spin $SU(2)$ case. One can easily check that these operators commute with the hamiltonian and the degeneracy of energy levels is enlarged to $(2S + 1)(2Q + 1)$, where S is the usual spin and Q the pseudo-spin. This extra symmetry will be used to show in the next section that the Bethe-ansatz states give the correct number of eigen-states of the 1D Hubbard model.

I now write the eigen-states of the hamiltonian in the following second quantized form

$$|\Psi\rangle_{N,M} = \sum_{\{1 \leq n_i \leq N_a\}} \phi(n_1, \ldots, n_M, n_{M+1}, \ldots, n_N) \prod_{j=1}^M c_{n_j,\downarrow}^\dagger \prod_{k=M+1}^N c_{n_j,\uparrow}^\dagger |0\rangle, \qquad (3.5)$$

where $n_1, \ldots, n_M$ are the positions of the M down spins, N the total number of electrons and ϕ the amplitude function. The anti-commutation relations among the creation operators eliminate all the terms with two up (or two down) spin electrons at the same site.

I now introduce a sector D_Q [25] defined as a region in the N particle space such that

$$D_Q := \{1 \leq n_{Q_1} < n_{Q_2} < \cdots < n_{Q_N} \leq N\}, \qquad (3.6)$$

where Q denotes a permutation of N numbers $\{1, 2, \ldots, N\}$. I also define a neighboring sector of D_Q as $D_{Q(ij)}$, where $j = i + 1$ and $Q(ij)$ is a permutation obtained from Q by interchanging Q_i and Q_j. Let ϕ_Q be the amplitude function in the sector D_Q and introduce the following ansatz for ϕ_Q

$$\phi_Q(n_1, \ldots, n_N) = \sum_P [Q, P] \exp(i \sum_j k_{P_j} n_{Q_j}). \tag{3.7}$$

P denotes a permutation of $\{1, \ldots, N\}$ associated with the quasi-momenta $\{k_i\}$ and Q with the electron site locations $\{n_i\}$. $[Q, P]$ is a coefficient that can be determined essentially from the boundary conditions imposed on the amplitude ϕ. Since ϕ_Q is a smooth function, I should have $\phi_Q = \phi_{Q(ij)}$ at the intersection of the two sectors (i.e. at $n_{Q_i} = n_{Q_j} = n$). From this continuity condition I obtain the following relationship,

$$[Q, P] + [Q, P(ij)] = [Q(ij), P] + [Q(ij), P(ij)]. \tag{3.8}$$

I start with the Schrödinger equation $H|\Psi\rangle = E|\Psi\rangle$, and obtain the following eigenvalue equation for ϕ

$$-\sum_{i=1}^{N} \sum_{s=\pm 1} \phi(n_1, \cdots, n_i + s, \cdots, n_N) \; + \; U\sum_{i>j} \delta(n_i - n_j)\phi(n_1, \cdots, n_N)$$
$$= E'\phi(n_1, \cdots, n_N), \tag{3.9}$$

where $E' = E + U(2N - N_a)/4$. Since no particles coincide in the sector D_Q, the on-site interaction term is identically zero and ϕ_Q satisfies

$$-\sum_{i=1}^{N} \sum_{s=\pm 1} \phi_Q(n_1, \cdots, n_i + s, \cdots, n_N) = E'\phi_Q(n_1, \cdots, n_N). \tag{3.10}$$

However, at the intersection between the two sectors D_Q and $D_{Q(ij)}$ (i.e. at $n_{Q_i} = n_{Q_j} = n$) the eigenvalue equation (3.9) is

$$\cdots + \phi(\cdots, n + 1, n, \cdots) + \phi(\cdots, n, n - 1, \cdots) + \phi(\cdots, n, n + 1, \cdots) +$$
$$\phi(\cdots, n - 1, n, \cdots) + \cdots + U\phi(\cdots, n, n, \cdots) = E'\phi(\cdots, n, n, \cdots), \tag{3.11}$$

where the obvious terms are omitted and are represented by dots. Note that the first and the second terms shown on the left side of the Eq. (3.11) necessarily belong to the sector $D_{Q(ij)}$ while the second and the forth terms to D_Q. I now evaluate Eq. (3.10) at $n_{Q_i} = n_{Q_j} = n$ (I can do this since ϕ is smooth), subtract it from Eq. (3.11) and use Eq. (3.8) to reduce the resulting expression to the following

$$[Q, P] = Y_{nm}^{ab}[Q, P'], \tag{3.12}$$

where

$$x_{nm} = -\frac{iU/2}{\sin k_n - \sin k_m + iU/2}, \tag{3.13}$$

$$Y_{nm}^{ab} = x_{nm} + (x_{nm} + 1)P^{ab}. \tag{3.14}$$

P^{ab} is the permutation operator which exchanges $Q_i = a$ and $Q_j = b$, and $P' = P(ij)$.
The operator Y satisfies the following consistency conditions

$$Y_{nm}^{ab} Y_{mn}^{ab} = 1, \tag{3.15}$$

$$Y_{ml}^{ab} Y_{nl}^{bc} Y_{nm}^{ab} = Y_{nm}^{bc} Y_{nl}^{ab} Y_{ml}^{bc}. \tag{3.16}$$

The above consistency conditions first obtained by C.N. Yang [109] for the 1D system
of particles with δ-function interaction is crucial to the integrability of this model and
becomes the basis for what is now known as the quantum Yang-Baxter equation [57].

By imposing the periodic boundary condition Eq. (3.12) has first been solved by
Yang [109] and later used by Lieb and Wu [77] to solve the 1D Hubbard model. In
particular Lieb and Wu obtained the following Bethe-ansatz equations:

$$\exp(ik_j N_a) = \prod_{\alpha=1}^{M} \frac{\sin k_j - \Lambda_\alpha + \frac{U}{4}i}{\sin k_j - \Lambda_\alpha - \frac{U}{4}i}, \tag{3.17}$$

$$\prod_{j=1}^{N} \frac{\Lambda_\alpha - \sin k_j + \frac{U}{4}i}{\Lambda_\alpha - \sin k_j - \frac{U}{4}i} = -\prod_{\beta=1}^{M} \frac{\Lambda_\alpha - \Lambda_\beta + \frac{U}{2}i}{\Lambda_\alpha - \Lambda_\beta - \frac{U}{2}i}, \tag{3.18}$$

where M and N_a are the numbers of down spin electrons and of sites, respectively,
and Λ_α is the spin rapidity introduced for each down spin. Following the notations
used by Woynarovich [106], I write the coefficient $[Q, P]$ as

$$[Q, P] = (-1)^Q (-1)^P \sum_R A(\Lambda_{R_1}, \Lambda_{R_2}, \ldots, \Lambda_{R_M}) \prod_{i=1}^{M} F_P(\Lambda_{R_i}; y_i), \tag{3.19}$$

where

$$F_P(\Lambda; y) = \frac{-i}{\sin k_{P_y} - \Lambda - i\frac{U}{4}} \prod_{j=1}^{y-1} \frac{\sin k_{P_j} - \Lambda + i\frac{U}{4}}{\sin k_{P_j} - \Lambda - i\frac{U}{4}}, \tag{3.20}$$

$$\frac{A(\ldots, \Lambda_{R_i}, \Lambda_{R_{i+1}}, \ldots)}{A(\ldots, \Lambda_{R_{i+1}}, \Lambda_{R_i}, \ldots)} = \frac{\Lambda_{R_{i+1}} - \Lambda_{R_i} + i\frac{U}{2}}{\Lambda_{R_i} - \Lambda_{R_{i+1}} + i\frac{U}{2}}. \tag{3.21}$$

The y's are unequal integers in increasing order and represent the down spin locations
in the series of spins in the sector D_Q. It has also been shown in [19] that the Bethe-
ansatz wavefunction (3.5) with $[Q, P]$ as given in Eq. (3.19) corresponds to the spin
and pseudo-spin highest weight state (i.e. $S^+|\Psi\rangle_{N,M} = Q^+|\Psi\rangle_{N,M} = 0$).

The eigen-energy and the crystal momentum are given by

$$E = \frac{U}{4}N_a - \sum_{j=1}^{N}(2\cos k_j + \frac{U}{2}), \tag{3.22}$$

$$P = \sum_{j=1}^{N} k_j. \tag{3.23}$$

The ground state is given by real k's and Λ's [77]. An excited state, however, is
described by the complex rapidities with a structure called "string" which is discussed
in the next section.

3.3 Numerical Demonstration of the Strings

I show in this section some of the numerical solutions of the Bethe-ansatz equations for the 1D Hubbard model and identify the strings in the solutions. First, let $z_j = \exp(ik_j)$ and turn the first Bethe-ansatz equation (3.17) into the following polynomial equation of order $N_a + 2M$ as in [39],

$$z_j^{N_a} \prod_{\alpha=1}^{M} (z_j^2 - (i\Lambda_\alpha - \frac{U}{2})z_j - 1) - \prod_{\alpha=1}^{M} (z_j^2 - (i\Lambda_\alpha + \frac{U}{2})z_j - 1) = 0. \qquad (3.24)$$

The $N_a + 2M$ roots for z_j are fixed by a given set of spin rapidities $\{\Lambda_\alpha\}$ and, hence, the spin rapidities fix the $N_a + 2M - N$ unoccupied as well as the N occupied pseudo-momenta. In the limit $U \to 0$ Eq. (3.24) becomes trivial to solve and factorizes to the following form

$$(z_j^{N_a} - 1) \prod_{\alpha=1}^{M} (z_j^2 - i\Lambda_\alpha z_j - 1) = 0. \qquad (3.25)$$

Therefore, in this free fermion limit there are N_a roots given by $z^{N_a} = 1$ and $2M$ roots associated with M Λ's. When the roots from the second factor of Eq. (3.25) are written as $(z - z^{-1})/2i$ (i.e. $\sin k$), the two roots associated with each Λ_α coincide. From the numerical inspection I find that when U is non-zero the two roots written in the form $\sin k$ associated with Λ_α splits up by an equal amount along the complex axis. The $2M$ roots associated with $\{\Lambda\}$ are called "complex" roots even though some of them can be real for some complex Λ's. N_a roots when written as $\sin k$ are always real and are thus called "real" roots. To illustrate the structure of the solutions for Eq. (3.24) I plot them in Fig. (3.1) for an arbitrary chosen set of Λ $\{-2, -1, 0, 1, 2\}$ at various values of U for $N_a = 10$ and $M = 5$.

Before I show the numerical result let us examine the simplest possible case, i.e. $N_a = N = 2$ and $M = 1$, where the complex roots can be observed. Two states obtained from the Bethe-ansatz equations are parameterized by the following rapidities: $\Lambda^{(\pm)} = 0$ and $\sin z^{(-)} = \pm\sqrt{1 - (\phi^{(-)})^2}$, $\sin z^{(+)} = \pm i\sqrt{(\phi^{(+)})^2 - 1}$, where $\phi^{(\pm)} = (U \pm \sqrt{U^2 + 64})/8$. The $(-)$ state is described by the two real roots and the $(+)$ state by a complex conjugate pair which I call Λ' 1-string. Note that the two complex roots go to zero as $U \to 0$ and to $\pm iU/4$ as $U \to \infty$. One can check that the pairs are very close to $\pm iU/4$ for large range of U not too close to zero. The wavefunctions for the two states, $(+)$ and $(-)$, are given by

$$|\Psi\rangle_{2,1}^{(-)} = -\phi^{(-)}|\psi_1\rangle + |\psi_2\rangle, \qquad (3.26)$$

$$|\Psi\rangle_{2,1}^{(+)} = -\phi^{(+)}|\psi_1\rangle + |\psi_2\rangle, \qquad (3.27)$$

where $|\psi_1\rangle = (c_{1,\downarrow}^\dagger c_{1,\uparrow}^\dagger + c_{2,\downarrow}^\dagger c_{2,\uparrow}^\dagger)|0\rangle$ and $|\psi_2\rangle = (c_{1,\downarrow}^\dagger c_{2,\uparrow}^\dagger - c_{2,\downarrow}^\dagger c_{1,\uparrow}^\dagger)|0\rangle$. Since $\phi^- \to 0$ and $\phi^+ \to \infty$ as $U \to \infty$, in that limit the $(-)$ state is given mostly by $|\psi_2\rangle$ (the unpaired electron state) and the $(+)$ state by $|\psi_1\rangle$ (the paired electron state). When U is small

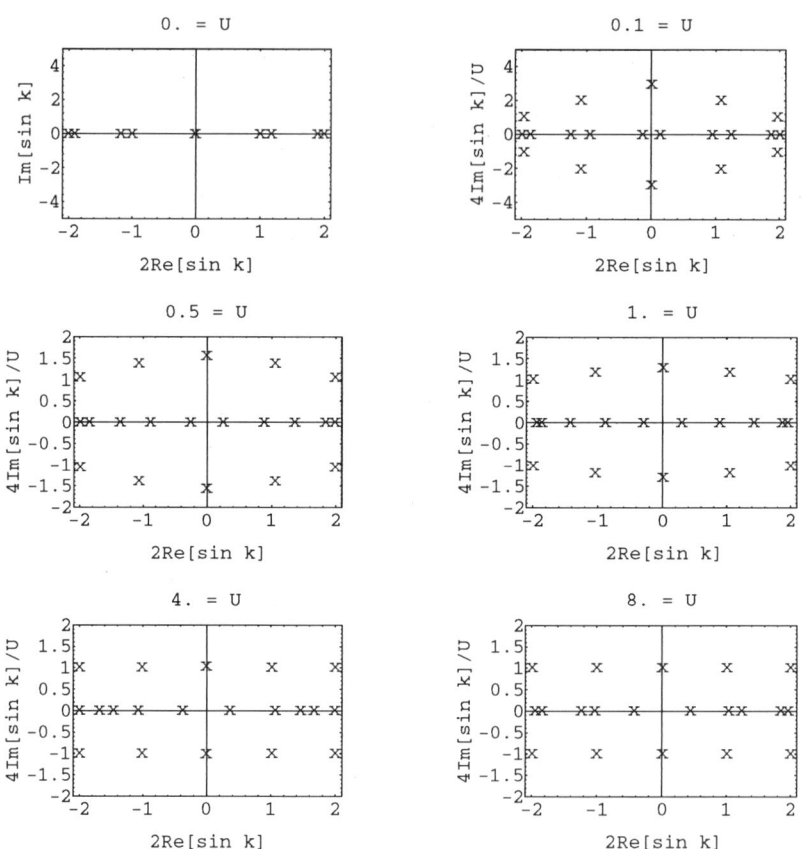

Fig. 3.1. Solutions of Eq. (3.24) for $\{\Lambda\} = \{-2, -1, 0, 1, 2\}$, $N_a = 10$, $M = 5$ and $U = 0, 0.1, 0.5, 1.0, 4.0, 8.0$. The horizontal and vertical axis correspond to $2\Re(\sin k)$ and $4\Im(\sin k)/U$, respectively, except for $U = 0$ in which case all the solutions are real and the vertical axis is arbitrarily chosen to be $\Im(\sin k)$. Note that for finite U the $2M$ "complex" and N_a "real" solutions separate out and as U is increased imaginary parts of the "complex" roots tend toward $\pm U/4$.

(i.e. comparable to the bandwidth $4t$), the real and complex roots do not correspond to the unpaired and paired electron states.

In order to solve Eq. (3.24) I need to find the spin rapidity set $\{\Lambda_\alpha\}$, and to determine $\{\Lambda\}$ I need $\{\sin k\}$. Thus, I have to start from an approximate solution for Λ's and use an iterative procedure to determine the correct $\{\Lambda\}$ and $\{\sin k\}$.

The second Bethe-ansatz equation (3.18) can be written in terms of $\lambda = 2\Lambda/U$ as follow

$$\prod_{j=1}^{N} \frac{\lambda_\alpha - 2(\sin k_j)/U + i/2}{\lambda_\alpha - 2(\sin k_j)/U - i/2} = - \prod_{\beta=1}^{M} \frac{\lambda_\alpha - \lambda_\beta + i}{\lambda_\alpha - \lambda_\beta - i}. \tag{3.28}$$

In Eq. (3.28) only the occupied pseudo-momenta are involved. If only the real roots are occupied and U very large, the Bethe-ansatz equation is reduced to that of the isotropy Heisenberg chain with N sites. Exact and complete set of numerical solutions of this equation for small sites (i.e. $N \leq 12$) has been worked out previously by Haldane [41]. Hence, I first solve Eq. (3.24) using the λ's for the Heisenberg chain and, then, choose a set of occupied roots which can be used to obtained a new set of λ's. I repeat this procedure until the roots do not change very much. A flow chart for this procedure is shown in Fig. 3.2.

The iterative procedure described above has been used to determine the correct values for the spin rapidities and the pseudo-momenta for the Hubbard model. In Fig. 3.3 I plot $\{2(\sin k)/U\}$ and $\{\lambda\}(= \{2\Lambda/U\})$ on the complex plane for $U = 10$ and $N_a = N = 6$. The λ's are represented by "+" and the occupied and unoccupied pseudo-momenta are represented by "x" and "o", respectively. Whenever there are two data points on top of each other I use the corresponding capital letters "X" and "O". Figure 3.3(a) corresponds to the singlet ground state with $N = 6$ and $M = 3$. In Fig. 3.3(b) the complex pair at the center is occupied and two holes on the real roots are created. There are some changes in the locations of all the other roots. The relative locations of the roots, however, are not changed; therefore, many different eigen-states corresponding to this root configuration can be obtained by selecting different sets of occupied roots. The root configuration shown in Fig. 3.3(c) has two roots at the origin. Hence, the corresponding pseudo-momenta would be 0 and π. In Fig. 3.3(d) there are one 2-string and one 1-string on top of each other (note the three "+" at $\pm 0.5i$ and 0). The two "x" at $\pm 0.5i$ correspond to the "+" at the origin. "O" at the origin represents two "o" associated with the other two "+". The root configurations shown in Fig. 3.3(e) and (f) are conjugate states where "x" for one state corresponds to "o" in the other state. The energies for the two such conjugate states can be shown to be equal and opposite [39]. I also checked the energies for each string configuration shown in Fig. 3.3 by the direct numerical diagonalization and found agreements to twelve significant figures.

Fig. 3.3(a)-(f) show that the string structure is quite well defined even for $N_a = 6$. I will show in the next section how one can systematically label the string states, count the states and get the thermodynamic equations.

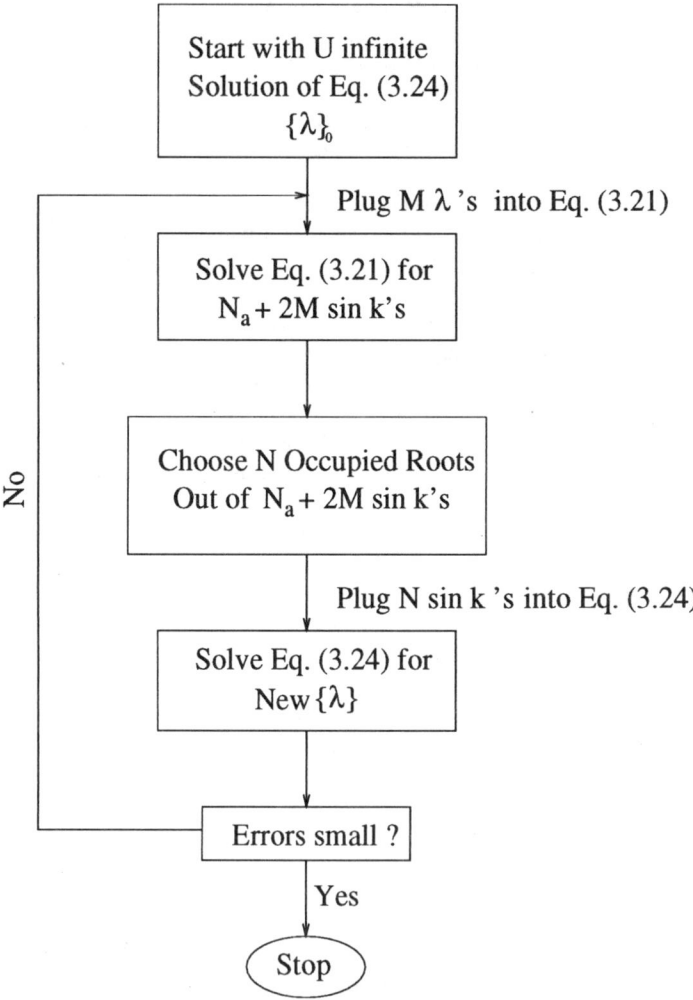

Fig. 3.2. A flow chart for iterative numerical procedure for solving the Bethe-ansatz equations for the Hubbard model. The initial set of rapidities $\{\lambda\}_0$ needed to start the procedure comes from solutions for the Heisenberg spin chain discussed in Chapter 2

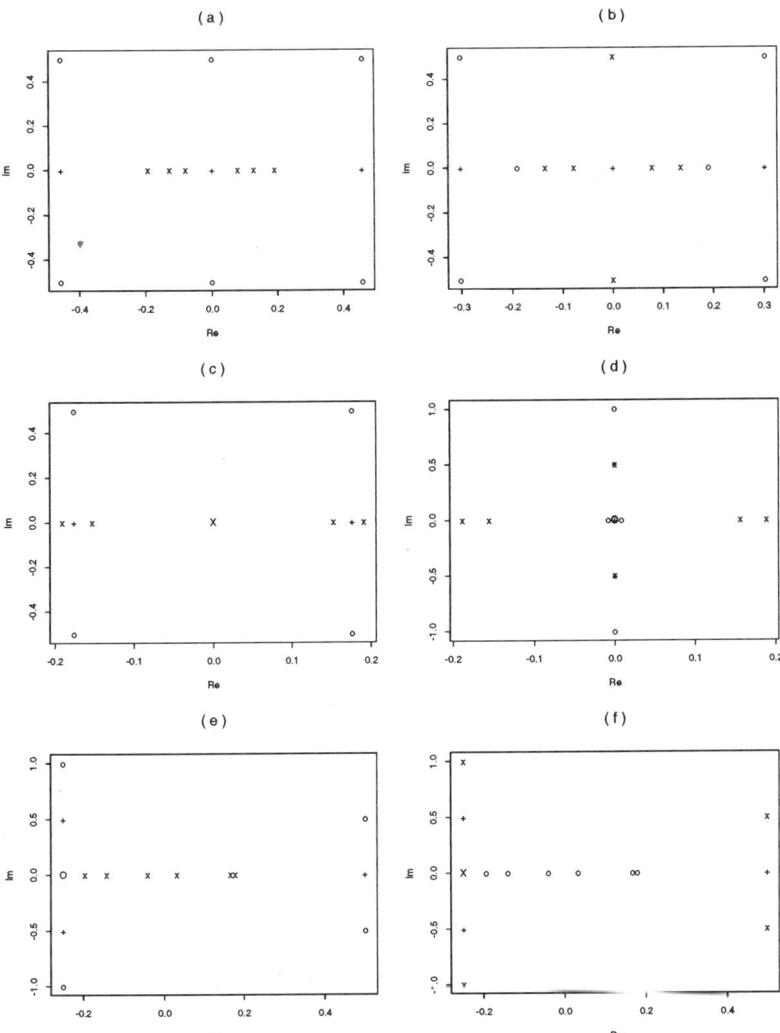

Fig. 3.3. $2 \sin k/U$ and $2\Lambda/U$ are plotted in the complex plane for $N_a, N = 6$ (i.e. $Q = 0$) and $U = 10.0$. "x" and "o" represent the occupied and unoccupied pseudo-momenta, respectively. Whenever two points are on top of each other, the corresponding capital letters, "X" and "O", are used. "+" represents the spin-rapidities. (a) The singlet ground state. $S = 0$ and $E = -16.6643627333$. (b) Same root configuration as in (a) with the central complex roots occupied. There are some changes in the relative locations of the roots. $S = 0$, $E = -4.9649086678$. (c) $S = 1$, $E = -16.3904965408$. Lowest excitation from the ground state. (d) 2-string and 1-string on top of each other at the center. $S = 0$, $E = -5.4064666563$. (e)(f) Two conjugate states where $E = \pm 15.7812788877$.

3.4 Thermodynamic Bethe-ansatz Equations

As discussed in previous section the following classification scheme for the general roots of the Bethe-ansatz equations (3.17) and (3.18) can be used:

1. k_j's represent the occupied real roots.

2. Complex k_j's (or Λ'-strings) corresponds to the occupied complex roots which are re-parameterized by the spin rapidities as follow

$$\sin k_{j,\alpha,n} = \Lambda'_{j,\alpha,n} \pm i\frac{U}{4}, \tag{3.29}$$

$$\Lambda'_{j,\alpha,n} = \Lambda'_{\alpha,n} + i\frac{U}{4}(n+1-2j), \tag{3.30}$$

where α is a label for the bound states of length n. $\Lambda'_{\alpha,n}$ is real, and $j = 1, 2, \ldots, n$. Λ' is used for the spin rapidities associated with the occupied complex roots and is distinguished from Λ.

3. Λ-strings are given by

$$\Lambda_{j,\alpha,n} = \Lambda_{\alpha,n} + i\frac{U}{4}(n+1-2j), \tag{3.31}$$

where $\Lambda_{\alpha,n}$ is real and $j = 1, 2, \ldots, n$. The complex roots associated with Λ's are unoccupied.

As seen from the numerical solution in the previous section, Eq. (3.29)-(3.31) are accurate description of the excited states even for a small system. When U is large compared to the bandwidth, the real k's correspond roughly to the unpaired electron states and the complex roots to the bound pairs.

After some algebraic manipulations the following equations for the real-k, Λ'-strings, and Λ-strings can be obtained:

$$k_j N_a = 2\pi I_j - \sum_{n=1}^{L} \sum_{\alpha=1}^{\nu_n} \theta\left(\frac{4(\sin k_j - \Lambda_\alpha^n)}{nU}\right) - \sum_{n=1}^{L'} \sum_{\alpha=1}^{\mu_n} \theta\left(\frac{4(\sin k_j - \Lambda_\alpha'^m)}{nU}\right) \tag{3.32}$$

$$N_a\left(\sin^{-1}(\Lambda_\alpha'^m + i\frac{U}{4}n) + \sin^{-1}(\Lambda_\alpha'^m - i\frac{U}{4}n)\right) = 2\pi J_\alpha'^m$$
$$+ \sum_{j=1}^{N-2L'} \theta\left(\frac{4(\Lambda_\alpha'^m - \sin k_j)}{nU}\right) + \sum_{m,\beta} \Theta_{nm}\left(\frac{4(\Lambda_\alpha'^m - \Lambda_\beta'^m)}{4U}\right), \tag{3.33}$$

$$\sum_{j=1}^{N-2L'} \theta\left(\frac{4(\Lambda_\alpha^n - \sin k_j)}{nU}\right) = 2\pi J_\alpha^n + \sum_{m,\beta} \Theta_{nm}\left(\frac{4(\Lambda_\alpha^n - \Lambda_\beta^m)}{U}\right), \tag{3.34}$$

where $\theta(x)$ and $\Theta_{nm}(x)$ are given by

$$\theta(x) = 2\mathrm{Arctan}\, x,$$

$$\Theta_{nm}(x) = \begin{cases} \theta(\frac{x}{|n-m|}) + 2\theta(\frac{x}{|n-m|+2}) + \cdots + 2\theta\left(\frac{x}{n+m-2}\right) + \theta(\frac{x}{n+m}) & \\ \qquad\qquad\qquad\qquad\qquad\qquad\qquad\qquad\qquad \text{if } n \neq m, & \\ 2\theta(\frac{x}{2}) + \cdots + 2\theta(\frac{x}{2n-2}) + \theta(\frac{x}{2n}) \qquad \text{if } n = m. & \end{cases} \tag{3.35}$$

ν_n and μ_n are the numbers of n-strings for Λ' and Λ, respectively, and $L' = \sum_n n\nu_n$ and $L = \sum_n n\mu_n$ with $L' + L = M$.

The quantum numbers for the Λ' and Λ n-strings, J'^m and J^n, satisfy the following conditions

$$|J'^m| \leq \frac{1}{2}(N_a - N + 2L' - \sum_{m=1}^{L'} c_{nm}\nu_m - 1), \qquad (3.36)$$

$$|J^n| \leq \frac{1}{2}(N - 2L - \sum_{m=1}^{L} c_{nm}\mu_m - 1), \qquad (3.37)$$

where $c_{nm} = 2\text{Min}(n,m) - \delta_{nm}$. Hence, the total number of partitioning the M spin rapidities into strings made up of Λ' (L' of them) and Λ (L of them) is

$$n_s(N_a, N, M, L, L') = \prod_{n=1}^{L'} \binom{N_a - N + 2L' - \sum_{m=1}^{L'} c_{nm}\nu_m}{\nu_n}$$
$$\times \prod_{k=1}^{L} \binom{N - 2L - \sum_{l=1}^{L} c_{kl}\mu_l}{\mu_k} \qquad (3.38)$$

L' spin rapidities Λ' represent the $2L'$ occupied complex pseudo-momenta. Therefore, there are only $N - 2L'$ real roots to choose from the N_a available real roots. The total number of solutions for fixed N_a, N and M is, thus, given by

$$n_t(N_a, N, M) = \sum_{L+L'=M} \binom{N_a}{N - 2L'} n_s(N_a, N, M, L, L'). \qquad (3.39)$$

The total degeneracy for the state due to the spin $SU(2)$ and the pseudo-spin $SU(2)$ symmetry is $(N_a - N + 1)(N - 2M + 1)$ as discussed in section 3.2. Hence, the total number of possible states associated with the Bethe-ansatz string solutions for a given N_a is

$$n_T(N_a) = \sum_{N=0}^{N_a} \sum_{M=0}^{[N/2]} (N_a - N + 1)(N - 2M + 1)n_t(N_a, N, M), \qquad (3.40)$$

where $[N/2]$ is the greatest integer $\leq N/2$. Since the pseudo-spin symmetry exists only for the even-site chain with the periodic boundary condition, Eq. (3.40) is meant for even N_a. I checked numerically that $n_T(N_a) = 4^{N_a}$ for $N_a \leq 12$. The Bethe-ansatz equations, therefore, give the correct number of the highest weight (spin and pseudo-spin) eigen-states. An analytical proof of the completeness of the string-states is given in [19].

For a large system one can introduce the distribution functions $\rho(k)$, $\sigma(\Lambda)$ and $\sigma'(\Lambda)$ for k, Λ_n and Λ'_n, respectively. The corresponding hole distribution functions are $\rho^h(k)$, $\sigma_n^h(\Lambda)$ and $\sigma_n'^h(\Lambda)$. By taking the continuum limit of Eq. (3.32)-(3.34) one obtains the following integral equations:

$$\frac{1}{2\pi} = \rho(k) + \rho^h(k) - \cos k \sum_{n=1}^{\infty} \int_{-\infty}^{+\infty} d\Lambda \, a_n(\Lambda - \sin k)(\sigma_n'(\Lambda) + \sigma_n(\Lambda)), \quad (3.41)$$

$$\int_{-\pi}^{+\pi} dk\rho(k)a_n(\Lambda - \sin k) = \sigma_n^h(\Lambda) + \sum_{m=1}^{\infty} \hat{A}_{nm}\sigma_m(\Lambda), \qquad (3.42)$$

$$\frac{1}{\pi}\Re\left(\frac{1}{\sqrt{1-(\Lambda-\frac{U}{4}i)^2}}\right) \; - \; \int_{-\pi}^{+\pi} dk\rho(k)a_n(\Lambda-\sin k) =$$

$$\sigma_n^{\prime h}(\Lambda) \; + \; \sum_{m=1}^{\infty}\hat{A}_{nm}\sigma_m^{\prime}(\Lambda), \tag{3.43}$$

where $\Re$ means "real part of," and $\hat{A}_{nm}$ is an operator defined by

$$\hat{A}_{nm}f(x) = \delta_{nm}f(x) + \frac{d}{dx}\int_{-\infty}^{+\infty}\frac{dx'}{2\pi}\,\Theta_{nm}\left(\frac{4(x-x')}{U}\right)f(x'), \tag{3.44}$$

The function a_n used in equations (3.41)-(3.43) is defined as

$$a_n(x) = \frac{1}{\pi}\frac{(U/4)n}{x^2+((U/4)n)^2}. \tag{3.45}$$

The operator $\hat{A}_n m$ and the function a_n are essentially same as those already appeared in Chapter 2.

The equilibrium distribution functions at temperature T can be obtained by minimizing the thermodynamic potential $\Omega = E - 2hS_z - A'N - TS$ with respect to the distribution functions subject to the conditions given in equations (3.41)-(3.43). This condition on Ω gives three equations in terms of $\zeta = \rho^h/\rho$, $\eta_n = \sigma_n^h/\sigma_n$ and $\eta_n' = \sigma_n^{\prime h}/\sigma_n'$:

$$\ln\zeta(k) \;=\; \frac{-2\cos k - h - A}{T}$$

$$+ \sum_{n=1}^{\infty}\int_{-\infty}^{+\infty}d\Lambda\, a_n(\Lambda-\sin k)\ln\{\frac{1+\eta_n'^{-1}(\Lambda)}{1+\eta_n^{-1}(\Lambda)}\}, \tag{3.46}$$

$$\ln(1+\eta_n(\Lambda)) \;=\; \frac{2nh}{T} - \int_{-\pi}^{+\pi}dk\,\cos k\, a_n(\Lambda-\sin k)\ln(1+\zeta^{-1}(k))$$

$$+ \sum_{m=1}^{\infty}\hat{A}_{nm}\ln(1+\eta_m^{-1}(\Lambda)), \tag{3.47}$$

$$\ln(1+\eta_n'(\Lambda)) \;=\; \frac{4\Re\left(\sqrt{1-(\Lambda-i\frac{U}{4}n)^2}\right) - 2nA}{T}$$

$$- \int_{-\pi}^{+\pi}dk\,\cos k\, a_n(\Lambda-\sin k)\ln(1+\zeta^{-1}(k))$$

$$+ \sum_{m=1}^{\infty}\hat{A}_{nm}\ln(1+\eta_m'^{-1}(\Lambda)), \tag{3.48}$$

where $A = A' + U/2$. The thermodynamic potential in terms of the newly defined distribution functions is given by

$$\frac{\Omega}{N_a} \;=\; 4\int_{-\infty}^{+\infty}d\Lambda\,\sigma_0(\Lambda)\,\Re\left(\sqrt{1-(\Lambda-i\frac{U}{4})^2}\right) - A$$

$$- T \int_{-\pi}^{+\pi} dk \, \rho_0(k) \ln(1 + \zeta^{-1}(k))$$

$$- T \int_{-\infty}^{+\infty} d\Lambda \, \sigma_0(\Lambda) \ln(1 + \eta_1'(\Lambda)), \tag{3.49}$$

where $\sigma_0(\Lambda)$ and $\rho_0(k)$ are the ground state distribution functions for half filled case and are given by

$$\sigma_0(\Lambda) = \int_{-\pi}^{+\pi} \frac{dk}{2\pi} \frac{1}{U} \text{sech}(\frac{4\pi(\Lambda - \sin k)}{U}), \tag{3.50}$$

$$\rho_0(k) = \frac{1}{2\pi} + \cos k \int_{-\infty}^{+\infty} d\Lambda \, a_1(\Lambda - \sin k)\sigma_0(\Lambda). \tag{3.51}$$

The equations (3.46)-(3.48) can be transformed into more useful forms. Let us define an operator $\hat{s}$ such that

$$\hat{s}f(\Lambda) = \int_{-\infty}^{+\infty} d\Lambda' \, \frac{1}{U} \text{sech}(\frac{2\pi(\Lambda - \Lambda')}{U}) f(\Lambda'). \tag{3.52}$$

Applying the operator $\hat{s}$ on the equations (3.47) and (3.48) one can obtain the expressions for $\ln(1 + \eta)$ and $\ln(1 + \eta')$. By substituting these expressions into (3.46) one obtains the following equation for ζ

$$\ln \zeta = -\frac{2 \cos k}{T} + \int_{-\infty}^{+\infty} d\Lambda \, \frac{1}{U} \text{sech}(\frac{2\pi(\Lambda - \sin k)}{U}) \times$$

$$\left[-\frac{4}{T} \Re \left(\sqrt{1 - (\Lambda - i\frac{U}{4})^2} \right) + \ln \left(\frac{1 + \eta_1'}{1 + \eta_1} \right) \right]. \tag{3.53}$$

By adding the expressions for $\ln(1 + \eta_{n-1})$ and $\ln(1 + \eta_{n+1})$ and applying $\hat{s}$ on the resulting expression one can obtain the following second order difference equation

$$\ln \eta_n = \hat{s} \ln\{(1 + \eta_{n-1})(1 + \eta_{n+1})\}, \tag{3.54}$$

where $n = 1, 2, 3, \cdots$. The boundary conditions for the difference equation are given by

$$\eta_0 = \exp\left\{-\int_{-\pi}^{+\pi} dk \, \cos k \, \delta(\Lambda - \sin k) \ln(1 + \zeta^{-1})\right\} - 1, \tag{3.55}$$

$$\lim_{n \to \infty} \frac{\ln \eta_n}{n} = 2z, \tag{3.56}$$

where $z = h/T$ and δ is a Dirac delta function. The equation for η' is exactly same as Eq. (3.54). The boundary conditions, however, are different and are given by

$$\ln \eta_n' = \hat{s} \ln\{(1 + \eta_{n-1}')(1 + \eta_{n+1}')\}, \tag{3.57}$$

$$\eta_0' = \exp\left\{-\int_{-\pi}^{+\pi} dk \, \cos k \, \delta(\Lambda - \sin k) \ln(1 + \zeta)\right\} - 1, \tag{3.58}$$

$$\lim_{n \to \infty} \frac{\ln \eta_n'}{n} = 2z', \tag{3.59}$$

where $z' = (U/2 - A)/T$. I assume that the chemical potential A is less than $U/2$ (i.e. less than half-filled Hubbard chain). The chain with more than half-filled band can readily be obtained by the canonical particle-hole transformation.

In order to obtain the thermodynamic potential one needs to solve the equations (3.54) and (3.57). These equations may be solved in a series form. The resulting expressions for η_1 and η_1' can then be substituted into (3.53) to obtain an expression for ζ. The equation for ζ is a transcendental equation and can be solved recursively. The thermodynamic potential can be obtained from Eq. (3.49) once the expressions for ζ and η_1' are found. Asymptotic solutions of the equations for η and η' are given in Appendix.

3.5   Calculation of the Thermodynamic Potential

In this section the thermodynamic potential is calculated using the asymptotic solutions for η and η'. By substituting the results given in the appendix into Eq. (3.53) I obtain

$$\ln(1 + \zeta^{-1}(k)) = \ln(1 + \zeta^{0-1}(k)) + S_1(\zeta(k)) + S_2(\zeta(k)) + \cdots, \tag{3.60}$$

where

$$\zeta^0 = e^{-(2\cos k + U/2)/T} \frac{\cosh z'}{\cosh z}, \tag{3.61}$$

$$S_1(\zeta(k)) = -\frac{1}{1 + \zeta^0(k)} F_1(\zeta(k)), \tag{3.62}$$

$$S_2(\zeta(k)) = \frac{\zeta^0(k)}{2(1 + \zeta^0(k))^2} F_1^2(\zeta(k)) - \frac{1}{1 + \zeta^0(k)} F_2(\zeta(k)), \tag{3.63}$$

where

$$F_1(\zeta(k)) = \frac{1}{U} \frac{1}{\cosh^2 z'} \left\{ \int_{-\pi}^{+\pi} \frac{dk}{2\pi} \cos k \, \ln(1 + \zeta^{-1}) - 1 \right\}$$
$$- \frac{1}{U} \frac{1}{\cosh^2 z} \int_{-\pi}^{+\pi} \frac{dk}{2\pi} \cos k \, \ln(1 + \zeta^{-1}) + O(\frac{1}{U^3}), \tag{3.64}$$

$$F_2(\zeta(k)) = \frac{3}{2} \frac{1}{U^2 T^2} \left\{ \frac{1 + 2\sinh^2 z'}{\cosh^4 z'} \left[T \int_{-\pi}^{+\pi} \frac{dk}{2\pi} \cos k \, \ln(1 + \zeta^{-1}) - 1 \right]^2 \right.$$
$$\left. - \frac{1 + 2\sinh^2 z}{\cosh^4 z} \left[T \int_{-\pi}^{+\pi} \frac{dk}{2\pi} \cos k \, \ln(1 + \zeta^{-1}) \right]^2 \right\} + O(\frac{1}{U^4}). \tag{3.65}$$

The expansion Eq. (3.60) is valid for $U \gg \sin k$.

I solve Eq. (3.60) recursively and substitute the resulting expression and the results given in the appendix into Eq. (3.49) to obtain

$$\frac{\Omega}{N_a} = \omega^{(0)} + \omega^{(1)} + \omega^{(2)} + \cdots, \tag{3.66}$$

where

$$\omega^{(0)} = \frac{U}{2} - A - T\ln(2\cosh z') - T\,I, \tag{3.67}$$

$$\omega^{(1)} = \frac{1}{U}\,J\left\{\left(\frac{1}{\cosh^2 z'} - \frac{1}{\cosh^2 z}\right)T\,I_c - \frac{1}{\cosh^2 z'}\right\}, \tag{3.68}$$

$$\begin{aligned}
\omega^{(2)} = & -\frac{1}{U^2}\,J\,J_c\left\{\left(\frac{1}{\cosh^2 z'} - \frac{1}{\cosh^2 z}\right)^2 T\,I_c \right. \\
& \left. -\frac{1}{\cosh^2 z'}\left(\frac{1}{\cosh^2 z'} - \frac{1}{\cosh^2 z}\right)\right\} \\
& +\frac{1}{U^2}\,J_c\,\frac{1}{\cosh^2 z'}\left\{\left(\frac{1}{\cosh^2 z'} - \frac{1}{\cosh^2 z}\right)T\,I_c - \frac{1}{\cosh^2 z'}\right\} \\
& +\frac{3}{2}\frac{1}{U^2 T}\,J\left\{\left(\frac{2}{\cosh^2 z'} - \frac{1}{\cosh^4 z'}\right)(T\,I_c - 1)^2 \right. \\
& \left. -\left(\frac{2}{\cosh^2 z} - \frac{1}{\cosh^4 z}\right)(T\,I_c)^2\right\} \\
& -\frac{1}{2}\frac{1}{U^2 T}\,K\left\{\left(\frac{1}{\cosh^2 z'} - \frac{1}{\cosh^2 z}\right)T\,I_c - \frac{1}{\cosh^2 z'}\right\}^2 \\
& -\frac{3}{2}\frac{1}{U^2 T}\left(\frac{2}{\cosh^2 z'} - \frac{1}{\cosh^4 z'}\right)(T\,I_c - 1)^2, \tag{3.69}
\end{aligned}$$

where

$$I = \int_{-\pi}^{+\pi}\frac{dk}{2\pi}\,\ln(1 + \zeta^{0-1}(k)), \tag{3.70}$$

$$I_c = \int_{-\pi}^{+\pi}\frac{dk}{2\pi}\,\cos k\,\ln(1 + \zeta^{0-1}(k)), \tag{3.71}$$

$$J = \int_{-\pi}^{+\pi}\frac{dk}{2\pi}\,\frac{1}{1 + \zeta^0(k)}, \tag{3.72}$$

$$J_c = \int_{-\pi}^{+\pi}\frac{dk}{2\pi}\,\frac{\cos k}{1 + \zeta^0(k)}, \tag{3.73}$$

$$K = \int_{-\pi}^{+\pi}\frac{dk}{2\pi}\,\frac{\zeta^0(k)}{(1 + \zeta^0(k))^2}. \tag{3.74}$$

The λ-expansion of $a_n\ln(1+\zeta^{-1})$ can be sorted in terms of $\frac{1}{U^n}f_n(\frac{1}{T},\frac{U}{T},\frac{A}{T})$ if $U \gg \sin k$. The expansion for Ω in Eq. (3.66) is this reorganized λ-expansion.

I now consider two different expansions which can be obtained from the reorganized λ-expansion: (i) $U,\ T \gg 1$ at fixed U/T and (ii) $U \gg 1$ at fixed T. The expansion (i) is given by $\sum O(\frac{1}{U^n T^m})$ and is a well defined power series. The expansion (ii) is given by $\sum O(\frac{1}{U^n}f_n(T))$ where $f_n(T)$ is a function of temperature. As long as $f_n(T)$ is not singular for some region of T the expansion is a well defined series in that region. I use the expansion (i) for the verification of the λ-expansion in the following section.

3.6   Verification of the Expansion for the Thermodynamic Potential

I show in this section of the chapter that the expansion for Ω agrees with the high temperature expansion worked out by Liu [78] whose work is based on the previous works by Kubo [69] and Brauneck [10]. In the limit $U, T \gg 1$ with U/T fixed the integrals in Eq. (3.71)-(3.74) can be expanded in $1/T$. I then obtain the following expansion for the thermodynamic potential

$$
\omega^{(0)} = -T \ln \xi - \frac{2}{T}\gamma(1 - 2\gamma) - \frac{1}{T^3}\gamma(\frac{1}{2} - 7\gamma + 24\gamma^2 - 24\gamma^3)
$$
$$
+ O(\frac{1}{T^5}), \tag{3.75}
$$
$$
\omega^{(1)} = -\frac{4}{U}\frac{\gamma^2}{\cosh^2 z}(1 - e^{-\frac{U}{T}})
$$
$$
- \frac{6}{UT^2}\frac{\gamma^2}{\cosh^2 z}(1 - 6\gamma + 8\gamma^2 + 2\gamma(1 - 4\gamma)e^{-\frac{U}{T}}) + O(\frac{1}{UT^4}), \tag{3.76}
$$
$$
\omega^{(2)} = -\frac{12}{U^2 T}\frac{\gamma^2}{\cosh^2 z}\left\{2\gamma + (1 - 2\gamma)e^{-\frac{U}{T}} - \frac{2\gamma^2}{\cosh^2 z}(1 - e^{-\frac{U}{T}})^2\right\}
$$
$$
+ O(\frac{1}{U^2 T^3}), \tag{3.77}
$$

where

$$
\xi = 1 + 2\cosh z \, e^{A/T} + e^{2A/T}e^{-U/T}, \tag{3.78}
$$
$$
\gamma = \frac{\cosh z \, e^{A/T}}{\xi}. \tag{3.79}
$$

Each term in the reorganized λ-expansion gives $\frac{1}{U^n}\sum_{m=0}^{\infty} O(\frac{1}{T^m})$. The terms of higher order which can be calculated readily are represented by the symbol O. The expressions in Eq. (3.75), (3.76) and (3.77) exactly agree with the expansion series given by Liu [78] who calculates them by the usual diagrammatic perturbation theory.

3.7   Investigation of Other Thermodynamic Quantities.

I obtain the density n, the charge susceptibility χ_c, the magnetic susceptibility χ_m (at constant A), and the specific heat C by differentiating the thermodynamic potential as

$$
n = -\frac{\partial \omega}{\partial A}, \quad \chi_c = \frac{\partial n}{\partial A},
$$
$$
\chi_m = -\frac{\partial^2 \omega}{\partial h^2}, \quad C = -T\frac{\partial^2 \omega}{\partial T^2}. \tag{3.80}
$$

These quantities are numerically calculated for $U = 8.0$ and plotted. Fig. 3.4(a)(b) show the density as a function of A and T, and Fig. 3.4(c) the chemical potential as a function of T at fixed densities. Note that at $n^* \approx 0.75$ the chemical potential is approximately independent of the temperature. Above n^* the chemical potential increases and below n^* it decreases as T is increased. This feature is due to emergence of the excitations of type 1, namely, the unpaired propagating electrons and holes (see Section 3.4) as the density is lowered. In other words the decrease in chemical potential A as a result of the decreasing number of charge-bound states is compensated by the increase in A with the rising number of excitations of type 1. Fig. 3.4(a) also shows that at $n = 1$ the chemical potential is fixed at $U/2$ due to appearance of the Hubbard gap.

In Fig. 3.4(a) the curve corresponding to $T = 0.05$ overshoots the value $n = 1$. Since our description cannot treat $n > 1$ case directly, this overshooting is an indication that the expansion up to the second order is not sufficient at this temperature. As discussed in appendix 3.A more than just first few terms in the λ-expansion for η and η' are required for temperatures below $O(1/U)$. The error is largest near $A = 2$ and disappears as $A \to U/2$. At $h > 0$ this problem goes away as shown in Fig. 3.4(b). Even at $T = 0.05$ the density curve remains strictly below the unity. The reason for this is that the expansion for the thermodynamic potential has a factor $[1/\cosh(h/T)]$ and this factor exponentially suppresses the other terms diverging as $1/T^m$ as $T \to 0$.

Figures 3.5(a),(b),(c) show temperature dependence of the charge susceptibility χ_c at fixed densities at $h = 0$, $h = 0.5$, and $h = 1.0$, respectively. χ_c diverges as $n \to 1$ and $T \to 0$, but at $n = 1$ χ_c vanishes as $T \to 0$. This behavior is due to the large charge fluctuations near the Hubbard gap. The density of states for a charged hole added to the half-filled Hubbard band diverges near the gap edge [65]. The diverging χ_c is a direct consequence of the diverging density of states of the charged hole. The charge susceptibility, furthermore, is almost independent of the magnetic field. This behavior suggests that the coupling between the charge and the spin is small, i.e. spin-charge separation at low temperatures.

The magnetic susceptibility χ_m (at constant A) is plotted in Fig. 3.6(a),(b),(c) at $h = 0$, $h = 0.5$, and $h = 1.0$, respectively. χ_m starts to decrease at certain temperatures in all three plots. At this temperature the system is beginning to favor the anti-ferromagnetic alignment. At $h = 0$ the magnetic susceptibility does not have any anomalous behavior [64]. Above certain magnetic field the magnetic susceptibility, however, shows the similar kind of anomaly as in the charge susceptibility as expected since $\chi_m \approx \chi_c$ at a large magnetic field. The susceptibility may also be defined as the derivative of $-\partial w/\partial h$ with respect to h at constant density which yields $\partial^2 w/\partial h^2 + (\partial^2 w/\partial A\partial h)^2/(\partial^2 w/\partial A^2)$. At $h = 0$ the two definitions are same since $(\partial^2 w/\partial A\partial h)|_{h=0} = 0$.

The specific heat as a function of temperature is shown in Fig. 3.7(a),(b),(c). At $h = 0$ the specific heat should have a hump due to the spin excitations at low temperature. At small temperatures our expansions start to break down as expected. The expansion, however, becomes better at low temperatures at a finite magnetic field. At $H = 0.5$ the hump due to the spin excitations is clearly shown in Fig. 3.7(b). Most of the spins line up with the magnetic field below a certain temperature. This polarization will suppress contributions from the spin excitations to the specific heat. At $h = 0.5$, this temperature is significantly below the most of the spin excitation spectrum. At $h = 1$, the temperature, however, is large enough so that the entire

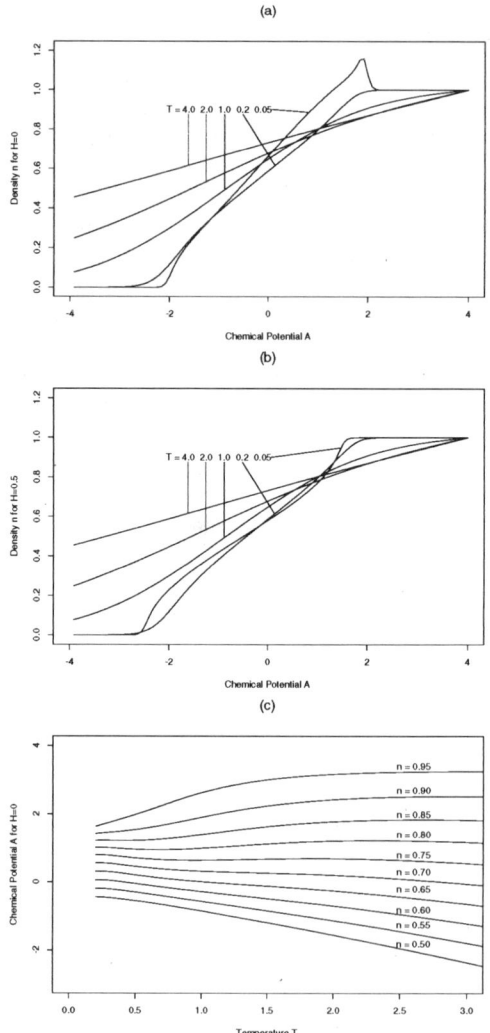

Fig. 3.4. (a) The density(n) is plotted against the the chemical potential(A) at fixed temperatures at zero magnetic field. The temperature, the chemical potential and the magnetic field are normalized with respect to t, the transfer integral. The curve corresponding to $T = 0.05$ overshoots the $n = 1$ value. This is an indication of the inadequacy of the second order expansion at low temperatures (i.e. $T < O(1/U)$). (b) n vs. A is plotted for $h = 0.5$. The overshooting is suppressed at this magnetic field. (c) The chemical potential as a function of the temperature at fixed densities is plotted. At $n^* \approx 0.75$ the chemical potential is approximately equal to 1 and is almost independent of the temperature.

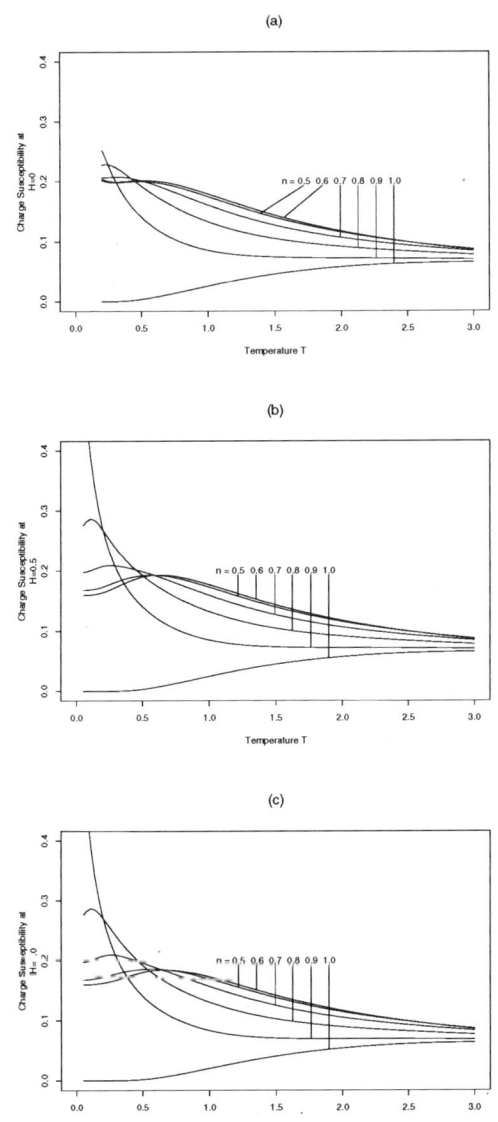

Fig. 3.5.   (a) The charge susceptibility or compressibility is plotted against T. As $n \rightarrow 1$ and $T \rightarrow 0$ the charge susceptibility diverges. At $n = 1$ the charge susceptibility, however, vanishes as $T \rightarrow 0$. (b) At h=0.5. (c) At h=1.0.

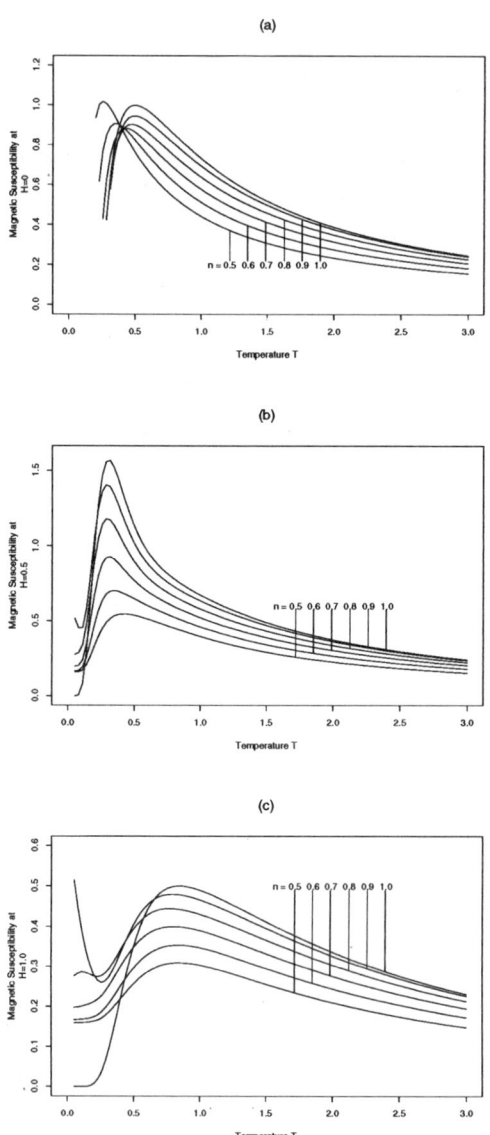

Fig. 3.6.   (a) Temperature-dependence of the magnetic susceptibility is shown at h=0. (b) At h=0.5. (c) At h=1.0 χ_m is beginning to resemble χ_c.

spin spectrum is suppressed as shown in Fig. 3.7(c).

The hump due to the excitations of the charge-bound state is clearly shown in Fig. 3.7(a),(b),(c) and it disappears as the density is lowered. When the chain is nearly half-filled, the excitations of charge-bound states become important at large enough temperature. This feature is consistent with existence of the gap in the charge-bound state spectrum [106]. At $n \approx 0.7$ this hump disappears and another hump at a lower temperature begins to appear. At $n = 0.6$ a distinct hump appears. This hump is due to excitations of the unpaired electrons and holes.

In Fig. 3.8(a),(b),(c) the specific heat at $h = 0.5$ is plotted for U=12.0, 15.0 and 20.0, respectively. The peak due to the spin excitation should appear at $T \approx 1/U$ and, therefore, should move to the left as U is increased. This behavior is not clearly indicated in the figures. It, however, is clear that the hump due to the complex k excitations move to the right as U increases. This behavior is simply due to the fact that as U is increased higher temperature is required to excite the charge pairs. Since the excitation due to the unpaired particles does not get directly affected by the magnitude of U, the real-k excitation peak does not move as U is changed. I also observe that the real-k excitation peak begins to rise at a higher density as U increases.

3.8 Conclusion

I show in this chapter some details of the string structure of the 1D Hubbard model and solve the thermodynamic Bethe-ansatz equations in a novel strong-coupling expansion series which is shown to be highly convergent and is consistent with the known high-temperature series expansion. Using this strong-coupling expansion various thermodynamic properties of 1D Hubbard model are studied and, in particular, the properties depend on three different types of excitations, namely, the spin-bound states (spin strings), the unpaired electrons and holes and the charge-bound states (charge strings). There also seems to be a characteristic density (for $U = 8$, $n^* \approx 0.7$) above which the excitations of charge-bound states are predominant and below which the unpaired electrons and holes are important. Some of these results are checked against the numerical works and found to agree. These agreements suggest that the string picture is valid in the thermodynamic limit.

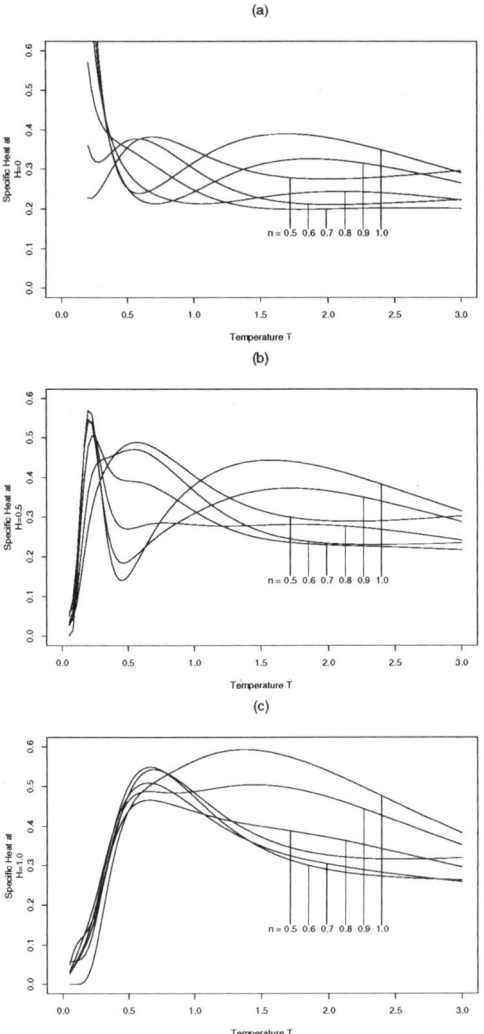

Fig. 3.7.   (a) Specific heat as a function of temperature at h=0. The three different types of excitations are clearly shown at three different temperatures. The spin excitations are at low T. The excitations of the unpaired electrons and holes are at intermediate T and at the densities smaller that 0.75. The charge-bound state excitations appear at the densities above 0.75 and at high T. (b) At h=0.5 the hump due to the spin excitations is clearly shown. (c) At h=1.0 the spin excitations are suppressed. The low temperature specific heat, hence, is due to the excitations of unpaired electrons and holes.

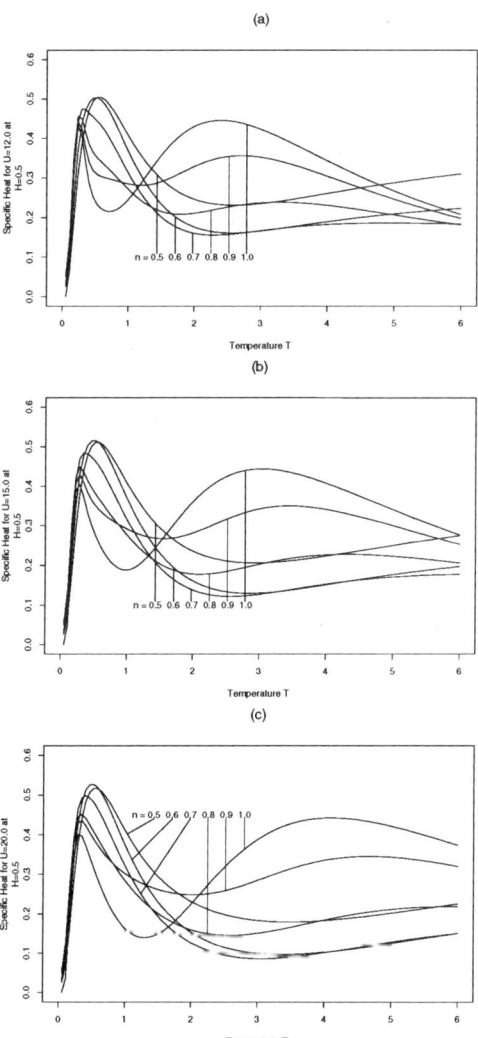

Fig. 3.8. Specific heat at $h = 0.5$ is plotted as a function of T for (a) $U = 12.0$, (b) $U = 15.0$ and (c) $U = 20.0$. It is difficult to tell whether the location of the peak due to the spin excitation moves as $1/U$. It, however, is clear that the hump due to the charge bound state excitation moves to the right as U increases. And, as expected the location of the peak due to the single particle and hole excitation remains the same as U changes.

3.A   How to Solve the Equations for η and η'.

In this appendix I show that the equations (3.54) and (3.57) can be solved if the integral terms in Eq. (3.55) and (3.58) are small. First, I introduce a parameter λ to represent the order of magnitude of the integral term and expand η_0 as

$$
\begin{aligned}
\eta_0 &= -\lambda \int_{-\pi}^{+\pi} dk \, \cos k \, \delta(\Lambda - \sin k) \ln(1 + \zeta^{-1}) \\
&+ \frac{1}{2!}\lambda^2 \left\{ \int_{-\pi}^{+\pi} dk \, \cos k \, \delta(\Lambda - \sin k)\ln(1 + \zeta^{-1}) \right\}^2 \\
&- \frac{1}{3!}\lambda^3 \left\{ \int_{-\pi}^{+\pi} dk \, \cos k \, \delta(\Lambda - \sin k)\ln(1 + \zeta^{-1}) \right\}^3 \\
&+ \cdots.
\end{aligned}
\tag{3.A.81}
$$

I then generalize the expansion to η_n as

$$
\eta_n = \eta_n^{(0)} + \lambda \eta_n^{(1)} + \lambda^2 \eta_n^{(2)} + \lambda^3 \eta_n^{(3)} + \cdots.
\tag{3.A.82}
$$

The expansion for η' is given by Eq. (3.A.81) and (3.A.82) with $\ln(1 + \zeta^{-1})$ replaced by $\ln(1 + \zeta)$. At first glance one might think that a "good" expansion for η means a "bad" expansion for η' and vice versa, since $\ln(1 + \zeta^{-1}) \ll 1$ necessitates $\ln(1 + \zeta) \gg 1$. However, that is not the case. Using the expression for $\ln \zeta$ given in Eq. (3.53) one can easily show that

$$
\begin{aligned}
\int_{-\pi}^{+\pi} dk \, \cos k \, \delta(\Lambda - \sin k)\ln \zeta &= -\frac{2}{T} \int dk \, \cos^2 k \, \delta(\Lambda - \sin k) \\
&= -\frac{4\Re(\sqrt{1 - \Lambda^2})}{T}
\end{aligned}
\tag{3.A.83}
$$

Only the first term in Eq. (3.53) contributes to the integral. Since $\ln(1 + \zeta) = \ln(1 + \zeta^{-1}) + \ln \zeta$, the difference between the expansions for η and η' can be ignored at high temperatures. And, for $|\Lambda| > 1$ $\eta_0 = \eta_0'$. One can also show that $\ln(1 + \zeta^{-1}) \to O(1/T)$ as $T \to 0$. Hence, as $T \to 0$ each term in the two expansions will both diverge as $1/T^m$.

One can determine $\eta_n^{(m)}$ by substituting Eq. (3.A.82) into Eq. (3.54)-(3.56), sorting the resulting expressions for each order and solving the equation for each order with the corresponding set of boundary conditions. The zeroth order equation and its boundary conditions are

$$
\eta_n^{(0)^2} = (1 + \eta_{n-1}^{(0)})(1 + \eta_{n+1}^{(0)}),
\tag{3.A.84}
$$

$$
\eta_0^{(0)} = 0,
\tag{3.A.85}
$$

$$
\lim_{n \to \infty} \frac{\ln \eta_n^{(0)}}{n} = 2z,
\tag{3.A.86}
$$

$\eta_n^{(0)}$ is independent of Λ. The operator $\hat{s}$, hence, becomes just a multiplicative factor $1/2$ and the solution to the zeroth order equation is given by

$$\eta_n^{(0)} = \left\{ \frac{\sinh(n+1)z}{\sinh z} \right\}^2 - 1. \tag{3.A.87}$$

The m-th order equation can be written in general as

$$\hat{s}\Upsilon_{n+1}^{(m)} - \frac{\sinh^2(n+1)z}{\sinh nz \, \sinh(n+2)z} \Upsilon_n^{(m)} + \hat{s}\Upsilon_{n-1}^{(m)} = Z_n^{(m)}. \tag{3.A.88}$$

The first order equation is given by Eq. (3.A.88) with

$$\Upsilon_n^{(1)} = \frac{\eta_n^{(1)}}{1 + \eta_n^{(0)}}, \tag{3.A.89}$$

$$Z_n^{(1)} = 0. \tag{3.A.90}$$

The first order equation is thus a homogeneous difference equation. The two corresponding boundary conditions are

$$\Upsilon_0^{(1)} = - \int_{-\pi}^{+\pi} dk \, \cos k \, \delta(\Lambda - \sin k) \ln(1 + \zeta^{-1}), \tag{3.A.91}$$

$$\lim_{n \to \infty} \frac{\Upsilon_n^{(1)}}{n} = 0. \tag{3.A.92}$$

The second order equation is given by Eq. (3.A.88) with

$$\Upsilon_n^{(2)} = \frac{\eta_n^{(2)}}{1 + \eta_n^{(0)}} - \frac{1}{2}\left(\frac{\eta_n^{(1)}}{1 + \eta_n^{(0)}}\right)^2, \tag{3.A.93}$$

$$Z_n^{(2)} = - \frac{\eta_n^{(1)2}}{2\eta_n^{(0)2}(1 + \eta_n^{(0)})}. \tag{3.A.94}$$

The corresponding boundary conditions are

$$\Upsilon_0^{(2)} = 0, \tag{3.A.95}$$

$$\lim_{n \to \infty} \frac{\Upsilon_n^{(2)}}{n} = 0. \tag{3.A.96}$$

The second and higher order equations are inhomogeneous equations with $Z_n^{(m)}$ being a function of $\eta_n^{(m-1)}$ and lower order η 's. The boundary conditions for higher order equations are still given by Eq. (3.A.95) and (3.A.96).

In order to solve the equation (3.A.88) I introduce the following Fourier transform

$$\tilde{\Upsilon}_n(\omega) = \int_{-\infty}^{+\infty} d\Lambda \, \Upsilon_n(\Lambda) e^{i\omega\Lambda}. \tag{3.A.97}$$

After the transformation the equation (3.A.88) becomes

$$\tilde{\Upsilon}_{n+2}^{(m)} - 2\cosh\frac{U}{4}\omega\frac{\sinh^2(n+2)z}{\sinh(n+1)z\,\sinh(n+3)z}\tilde{\Upsilon}_{n+1}^{(m)} + \tilde{\Upsilon}_n^{(m)} = \tilde{Q}_n^{(m)}, \qquad (3.A.98)$$

where $\tilde{Q}_n^{(m)} = 2\cosh\frac{U}{4}\omega\,\tilde{Z}_{n+1}^{(m)}$, and n is shifted by one. The solution to the homogeneous part of the equation is given by

$$\tilde{\Upsilon}_n^{h(m)} = A^+(\omega)\tilde{\Upsilon}_n^+ + A^-(\omega)\tilde{\Upsilon}_n^-, \qquad (3.A.99)$$

where $A^{\pm}$ are some functions of ω and

$$\tilde{\Upsilon}_n^{\pm} = \frac{\sinh nz}{\sinh(n+1)z}e^{\pm(n+2)\frac{U}{4}|\omega|} - \frac{\sinh(n+2)z}{\sinh(n+1)z}e^{\pm n\frac{U}{4}|\omega|}. \qquad (3.A.100)$$

The full solution to Eq. (3.A.98) is

$$\tilde{\Upsilon}_n^{(m)} = \tilde{\Upsilon}_n^{h(m)} + \tilde{\Upsilon}_n^-\sum_{j=0}^{n-1}\frac{\tilde{\Upsilon}_{j+1}^+}{K_0}\tilde{Q}_j^{(m)} - \tilde{\Upsilon}_n^+\sum_{j=0}^{n-1}\frac{\tilde{\Upsilon}_{j+1}^-}{K_0}\tilde{Q}_j^{(m)}, \qquad (3.A.101)$$

where $K_0 = \tilde{\Upsilon}_0^+\tilde{\Upsilon}_1^- - \tilde{\Upsilon}_0^-\tilde{\Upsilon}_1^+$.

By imposing the boundary conditions the constants can be determined. The solution to Eq. (3.A.98) for $m = 1$ is

$$\tilde{\Upsilon}_n^{(1)} = \frac{\sinh z}{\sinh 2z}\left\{\frac{\sinh nz}{\sinh(n+1)z}\int_{-\pi}^{+\pi}dk\,\cos k\,e^{i\omega\sin k}e^{-\frac{U}{4}(n+2)|\omega|}\ln(1+\zeta^{-1})\right.$$
$$\left. - \frac{\sinh(n+2)z}{\sinh(n+1)z}\int_{-\pi}^{+\pi}dk\,\cos k\,e^{i\omega\sin k}e^{-\frac{U}{4}n|\omega|}\ln(1+\zeta^{-1})\right\}. \qquad (3.A.102)$$

The solution to Eq. (3.A.98) for $m = 2$ and $n = 1$ is

$$\tilde{\Upsilon}_1^{(2)} = \frac{\cosh\frac{U}{4}\omega}{\tilde{\Upsilon}_0^+}\sum_{j=0}^{\infty}\left\{\frac{(1+\eta_{j+1}^{(0)})\tilde{\Upsilon}_{j+1}^-}{(\eta_{j+1}^{(0)})^2}\right\}(\widetilde{\Upsilon_{j+1}^{(1)}})^2. \qquad (3.A.103)$$

The solutions for η' are also given by Eq. (3.A.102) and (3.A.103) with z and ζ^{-1} replaced by z' and ζ.

By inverse Fourier transforming Eq. (3.A.101) and reorganizing some terms I find the followings

$$\Upsilon_n^{(1)} = -G + \frac{1}{4\cosh^2 z}\Delta G, \qquad (3.A.104)$$

$$\Upsilon_n'^{(1)} = -G + \frac{4}{T}\left\{\Re\sqrt{1-(\Lambda-i\frac{U}{4})^2} - \frac{U}{4}\right\}$$
$$+ \frac{1}{4\cosh^2 z'}\left\{\Delta G - \int_{-\pi}^{+\pi}dk\frac{2\cos^2 k}{T}\Delta a\right\}, \qquad (3.A.105)$$

where

$$G = \int_{-\pi}^{+\pi} dk \, \cos k \, a_1(\Lambda - \sin k) \ln(1 + \zeta^{-1}), \qquad (3.A.106)$$

$$\Delta G = \int_{-\pi}^{+\pi} dk \, \cos k \, \Delta a \, \ln(1 + \zeta^{-1}), \qquad (3.A.107)$$

$$\Delta a = \int_{-\infty}^{+\infty} \frac{d\omega}{2\pi} \, 2\cosh\frac{U}{4}\omega \, e^{-\frac{U}{2}|\omega| + i\omega(\Lambda - \sin k)}. \qquad (3.A.108)$$

In calculating the next order I encounter integrals with kernel $a_2(\sin k - \sin k')$. I expand the kernel as $a_2 = [2/(\pi U)]\{1 - [2(\sin k - \sin k')/U]^2 + \cdots\}$. If I restrict our calculation to $O(1/U^2)$, I may approximate the kernel as $a_n \approx 4/(n\pi U)$. This approximation makes the subsequent calculations simple and, especially, the evaluation of the infinite sum in Eq. (3.A.103) much easier. The results are

$$\int_{-\infty}^{+\infty} d\Lambda \, \frac{1}{U} \, \text{sech}\frac{2\pi(\Lambda - \sin k)}{U} \Upsilon_1^{(2)}(\Lambda) =$$
$$\frac{3}{2} \frac{1}{U^2 T^2} \frac{1 + 2\sinh^2 z}{\cosh^4 z} \left\{ T \int_{-\pi}^{+\pi} \frac{dk}{2\pi} \, \cos k \, \ln(1 + \zeta^{-1}) \right\}^2, \quad (3.A.109)$$

$$\int_{-\infty}^{+\infty} d\Lambda \, \frac{1}{U} \, \text{sech}\frac{2\pi(\Lambda - \sin k)}{U} \Upsilon_1'^{(2)}(\Lambda) =$$
$$\frac{3}{2} \frac{1}{U^2 T^2} \frac{1 + 2\sinh^2 z'}{\cosh^4 z'} \left\{ T \int_{-\pi}^{+\pi} \frac{dk}{2\pi} \, \cos k \, \ln(1 + \zeta^{-1}) - 1 \right\}^2 (3.A.110)$$

I do not need to know $\Upsilon_1^{(2)}(\Lambda)$ explicitly to calculate the thermodynamic potential: Eq. (3.A.109) and (3.A.110) will be sufficient.

As one can see in the calculations above nth order expansion term involve integrals with kernels a_n. The delta function in Eq. (3.A.81) can then be considered as the limit of a_n as defined in Eq. (3.45) with $n \to 0$, and nth term in Eq. (3.A.82) will be some functions of the integral $\int dk \, \cos k \, a_n(\Lambda - \sin k) \ln(1 + \zeta^{-1})$. Since $a_n \leq O(1/U)$, the mth order term in Eq. (3.A.82) will then be $O(1/(UT)^m)$ at low temperatures. I find this result in Eq. (3.A.109) and (3.A.110) for $z, z' - 0$. Hence, at temperatures smaller than $O(1/U)$ the second order term gets large compared to $O(1)$ and I need to calculate the higher order terms to approximate η and η' (for $z, z' = 0$). For $z, z' \neq 0$ the second order terms vanish exponentially as $T \to 0$. The higher order terms are expected to behave similarly.

Chapter 4

Models with Inverse-Square Exchange

4.1 Introduction

In the previous few chapters some of the Bethe-ansatz solvable models are discussed and, in some cases, extremely detailed calculations with heavy algebra are carried out for some less than fair amount of physical insight. Perhaps the single most important lesson one gets out of this class of models is that the complete excitation spectra are described by the strings that can be considered as quantum solitonic excitations with internal quantum degrees of freedom.[1] For some reason, however, the strings have been studied very little and not much more than what has been presented in the previous two chapters are known about them. It is probably because of the algebraic complexities involved. There is a much simpler system where these strings are present. It is known as the Moser-Calogero-Sutherland-Haldane-Shastry systems which have characteristic long-range, inverse-square interaction.

The string structure for $SU(n)$ spin chain will be discussed in the next chapter and in this chapter I introduce and show some exact treatments of the inverse-square exchanges both in continuum space and on lattice. [31] In particular a class of eigenstates of the hamiltonians is shown to be given by the Jastrow-product. It is also shown that some states of the continuum models survive the generalization to lattice depending on the density and the dimensionless interaction coupling constant λ. Generalizations of spinless Calogero-Sutherland system to $SU(n)$ spinful system are also shown. Figure 4.1 shows how various models with the inverse-square exchange have evolved from the original Calogero-Sutherland model. In the figure the models that are developed in [31] and presented in this chapter are boxed in ovals. The solid arrows mean "generalized from" or "evolved from." Two equivalent models that are represented in different ways are connected by a dashed arrow. It is emphasized that while the bosonic t-J model (λ odd) is a direct generalization of the bosonic supersymmetric case ($\lambda = 1$), the fermionic t-J model (λ even) is not directly generalizable from the fermionic supersymmetric t-J model.

---

[1] For systems with no internal degrees of freedom only the simplest 1-strings are present.

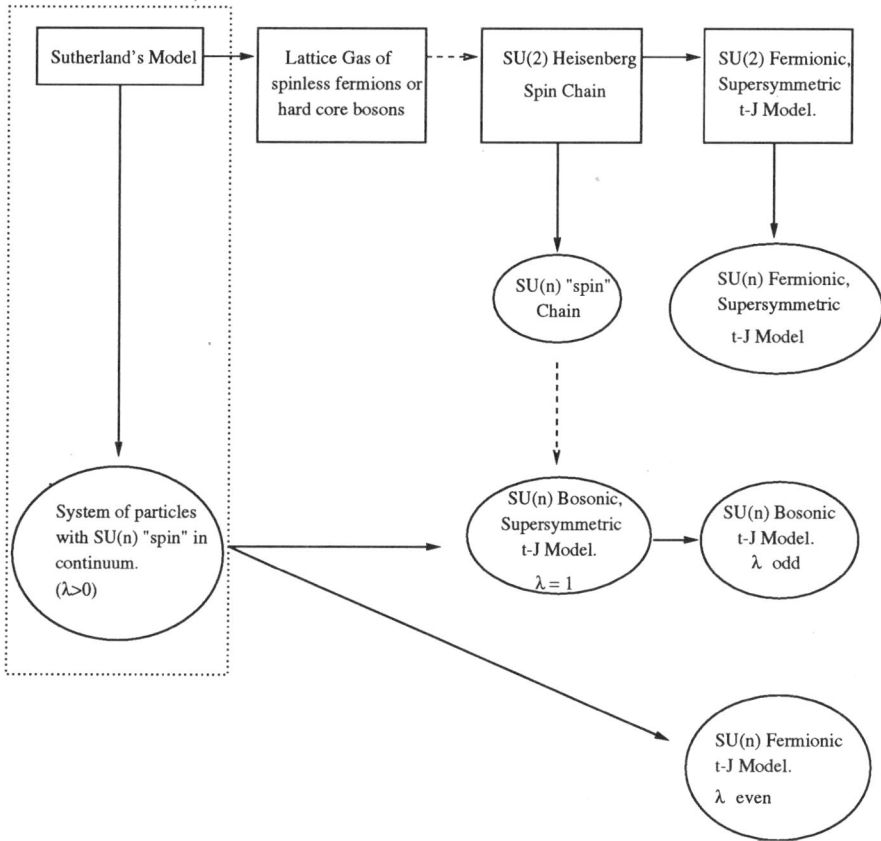

Fig. 4.1. Evolution of the inverse-square exchange models both in the continuum space and lattice. The models enclosed in the dotted box are continuum version and the rest of the models are lattice version. The solid arrows mean "evolved from" or "generalized from." The dashed arrows connect two equivalent models in disguise.

4.2 Calogero-Sutherland Model and Multicomponent Generalization

The Calogero-Sutherland model with $SU(n)$ internal spin degrees of freedom is studied in this section. For convenience let us consider a periodic system, where the hamiltonian in units of $\hbar^2/m$ is given by

$$H = -\frac{1}{2}\sum_i \frac{\partial^2}{\partial x_i^2} + \lambda \sum_{i<j} \frac{\lambda + P_{ij}^\sigma}{d(x_i - x_j)^2}, \qquad (4.1)$$

where λ is the dimensionless interaction parameter. P_{ij}^σ is an operator that exchanges particle spins at x_i and x_j, and $d(x) = (L/\pi)|\sin(\pi x/L)|$. $d(x_i - x_j)$ is the chord distance between particles at x_i and x_j on a circle with circumference L. If all particles have the same spin, this model reduces to the system of spinless particles studied by Sutherland. Note that Sutherland's coupling parameter λ' corresponds to $\lambda - 1$ in our notation. For example, the spinless free Fermi gas corresponds to $\lambda' = 1$, but in our notation $\lambda = 0$. In recent year there have been many advancements in understanding this spinless system and, thus, it deserves to be presented in a separate chapter. Let us concentrate on multicomponent generalizations of this class of model in this chapter.

First, it is easily found that the wavefunction Ψ must vanish as $|x_i - x_j|^{\lambda+1}$ $(|x_i - x_j|^\lambda)$ as $x_i - x_j \to 0$ in the case of symmetric (antisymmetric) spin configuration of the two particles at x_i and x_j. And if $0 < \lambda < 1$ the effective interaction strength is attractive for the antisymmetric spin configuration and there is some ambiguity in this case as it must be further specified whether Ψ vanishes as $|x_i - x_j|^\lambda$ or as $|x_i - x_j|^{1-\lambda}$ as particles approach. If we choose the first boundary condition the free fermion limit is obtained as $\lambda \to 0$. There are also two possible ways to interpret the interactions for this system. The first is to consider the system as a ring embedded in a plane; hence, even though the particles are constrained to move in one dimension, the interaction is two dimensional in nature. The other is to regard the system as strictly one dimensional by taking $1/d(x_i - x_j)^2$ as the effective interaction after summing the pairwise $1/x^2$ interaction around the ring infinite times as follow

$$\sum_{n=-\infty}^{+\infty} \frac{1}{(x + nL)^2} = \frac{1}{d(x)^2}. \qquad (4.2)$$

This feature is special to the inverse-square potential.

In analogy with the states previously constructed for the $SU(2)$ spin chain in Ref. [40], we propose the following Jastrow-product-type wavefunctions for our hamiltonian:

$$\Psi(\{z\sigma\}) = \Psi_0 \prod_k z_k^{J_{\sigma_k}}, \qquad (4.3)$$

where

$$\Psi_0 = \prod_{n>m} \phi_{nm},$$

$$g_{nm}(x) = \left(\frac{z_n - z_m}{|z_n - z_m|}\right)^x, \tag{4.4}$$

$$\phi_{nm} = g_{nm}(x)|z_n - z_m|^\lambda (z_n - z_m)^{\delta_{\sigma_n \sigma_m}} \exp[i\tfrac{1}{2}\pi \mathrm{sgn}(\sigma_n - \sigma_m)]. \tag{4.5}$$

Here, $z_n = \exp(2\pi i x_n / L)$, δ is the Kronecker delta function, x the statistical parameter, and σ_n is the ordered spin index and J_σ the global current of particles with spin σ. The currents J_σ are taken to be integers and are restricted to some allowed values which are given later in this section. Note also that the wavefunction with $\lambda = x = 0$ is the Slater determinant that corresponds to the states of free $SU(n)$ fermions.

Symmetry of the wavefunction with respect to exchange of particles is given as

$$\Psi(\ldots, z_i\sigma_i, \ldots, z_j\sigma_j, \ldots) = (-1)^{x+1}\Psi(\ldots, z_j\sigma_j, \ldots, z_i\sigma_i, \ldots). \tag{4.6}$$

Hence, x can be considered as the statistical parameter and $g(x)$ the corresponding singular gauge transformation.[2]

The total hamiltonian can be divided as $H = H^0 + H^1 + H^2$ where H^0, H^1, and H^2 are the kinetic, potential, and spin exchange hamiltonian, respectively. One can show that each operator acting on the wavefunction gives two types of terms, "wanted" and "unwanted." "Wanted" terms are defined to be the terms that depend only on the global variables such as the total number of particles, M_σ, J_σ, etc. "Unwanted" terms explicitly depend on the local variables z_i and σ_i. Since the eigen-energy should depend only on the global variables, the "unwanted" terms for H^0, H^1, and H^2 should cancel or combine to give "wanted" terms.

First let us examine H^0 acting on the wavefunction and define the following derivatives:

$$\varphi_{ij} \equiv \frac{\partial_{z_j}\phi_{ij}}{\phi_{ij}} = -\frac{\lambda}{z_i - z_j} - \frac{\lambda - x}{2z_j} - \frac{\delta_{\sigma_i \sigma_j}}{z_i - z_j}, \tag{4.7}$$

$$\xi_{ij} \equiv \partial_{z_j}\varphi_{ij}, \tag{4.8}$$

$$\eta_j^{(1)} \equiv z_j \frac{\partial_{z_j}\Psi_0}{\Psi_0} = z_j \sum_{i(\neq j)} \varphi_{ij}, \tag{4.9}$$

$$\eta_j^{(2)} \equiv z_j^2 \frac{\partial_{z_j}^2 \Psi_0}{\Psi_0} = z_j^2 \sum_{i(\neq j)} \xi_{ij} + (\eta_j^{(1)})^2. \tag{4.10}$$

In terms of these derivatives $H^0\Psi$ is given as

$$H^0\Psi = 2(\pi/L)^2 \sum_j [\eta_j^{(2)} + (2J_{\sigma_j} + 1)\eta_j^{(1)} + J_{\sigma_j}^2]\Psi, \tag{4.11}$$

and further calculations show that H^0 acting on the wavefunction gives "wanted" (W_K) and "unwanted" (U_K) terms which are given as follow

$$H^0\Psi = 2\left(\frac{\pi}{L}\right)^2 (W_K + U_K)\Psi, \tag{4.12}$$

---

[2]Refer to Chapter 7 for more details about the fractional statistics.

where

$$W_K = \frac{\lambda^2}{12} N(N^2 - 1) + \frac{1}{2}(\lambda + 1) \sum_\sigma M_\sigma(M_\sigma - 1) + \frac{1}{3} \sum_\sigma M_\sigma(M_\sigma - 1)(M_\sigma - 2)$$
$$- \sum_\sigma M_\sigma J_\sigma(\lambda N - M_\sigma - J_\sigma - \lambda + 1) + E(x), \tag{4.13}$$

$$U_K = \lambda \sum_{i \neq j} \frac{z_i z_j}{(z_i - z_j)^2} \left\{ \lambda - 1 + 2\delta_{\sigma_i \sigma_j} \right\} + \lambda \sum_{i \neq j} \frac{z_i J_{\sigma_i} - z_j J_{\sigma_j}}{z_i - z_j}$$
$$+ \lambda \sum_{(i \neq j \neq k)} \frac{z_i(z_i + z_k)}{(z_i - z_j)(z_i - z_k)} \delta_{\sigma_i \sigma_j}, \tag{4.14}$$

where the statistical energy $E(x)$ is given by

$$E(x) = x^2 \frac{N(N-1)^2}{4} + x \frac{N-1}{2} \sum_\sigma M_\sigma(M_\sigma - 1) + x(N-1) \sum_\sigma M_\sigma J_\sigma. \tag{4.15}$$

Here, M_σ is the total number of σ particles in the system. One can now anticipate U_K to be canceled or combined with the "unwanted" terms from H^1 and H^2 to give local variable independent terms.

The potential energy H^1 acts trivially on Ψ to give $\lambda^2 \sum_{i<j} 1/d(x_i - x_j)^2$. The action of H^2 on the wavefunction is less trivial and one needs to evaluate the following expression

$$P_{\{M\}} \equiv \frac{1}{2\Psi} \sum_{i>j} \frac{P_{ij}^\sigma \Psi}{d(x_i - x_j)^2} = -\frac{1}{\Psi(\{z\sigma\})} \sum_{i \neq j} \frac{z_i z_j}{(z_i - z_j)^2} \Psi(\{z\sigma'\}), \tag{4.16}$$

where $\{z\sigma'\}$ is a configuration with σ_i and σ_j exchanged with respect to $\{z\sigma\}$, and likewise $\{z'\sigma\}$ is equal to $\{z\sigma\}$ with z_i and z_j exchanged. Using the following identity,

$$\Psi(\{z\sigma\}) = (-1)^{x+1} \Psi(\{z'\sigma'\}), \tag{4.17}$$

one may rewrite $P_{\{M\}}$ as

$$P_{\{M\}} = (-1)^{x+1} \sum_{i \neq j} \frac{z_i z_j}{(z_i - z_j)^2} \frac{\Psi(\{z\sigma'\})}{\Psi(\{z'\sigma'\})}$$
$$= \sum_{i \neq j} \frac{z_i z_j}{(z_i - z_j)^2} (-1)^{\delta_{\sigma_i \sigma_j}} \left(\frac{z_i}{z_j} \right)^{J_{\sigma_j} - J_{\sigma_i}} \prod_{k \neq ij} \left(\frac{z_k - z_i}{z_k - z_j} \right)^{\delta_{\sigma_j \sigma_k} - \delta_{\sigma_i \sigma_k}}. \tag{4.18}$$

The action of P^σ is independent of x and λ since it does not involve particle exchange or depend explicitly on the interaction parameter λ. In order to simplify P^σ the following identity is useful

$$\left(\frac{z}{z'} \right)^n \equiv \sum_{q=0}^{|n|} \binom{|n|}{q} \left\{ \left(\frac{z - z'}{z'} \right)^q \theta(n) + \left(\frac{z' - z}{z} \right)^q \theta(-n) \right\}, \tag{4.19}$$

where $\theta(n)$ is the step function with $\theta(0) \equiv 1/2$. Using this identity one can rewrite Eq. (4.18) as

$$P_{\{M\}} = P_0 + \sum_{\sigma,\sigma'} \sum_{q=1}^{|J_{\sigma'} - J_{\sigma}|} P_q^{\sigma\sigma'}. \tag{4.20}$$

where P_0 and $P_q^{\sigma\sigma'}$ are given by

$$P_0 = \sum_{i \neq j} \frac{z_i z_j}{(z_i - z_j)^2} \left\{ -\delta_{\sigma_i \sigma_j} + (1 - \delta_{\sigma_i \sigma_j}) \prod_{k \neq ij} \left(\frac{z_k - z_i}{z_k - z_j} \right)^{\delta_{\sigma_j \sigma_k} - \delta_{\sigma_i \sigma_k}} \right\} \tag{4.21}$$

$$\begin{aligned} P_q^{\sigma\sigma'} = {} & 2\sum_{i \neq j} \frac{z_i z_j}{(z_i - z_j)^2} \delta_{\sigma\sigma_i} \delta_{\sigma'\sigma_j} (1 - \delta_{\sigma\sigma'}) \prod_{k \neq ij} \left(\frac{z_k - z_i}{z_k - z_j} \right)^{\delta_{\sigma_j \sigma_k} - \delta_{\sigma_i \sigma_k}} \\ & \times \binom{J_{\sigma'} - J_\sigma}{q} \theta(J_{\sigma'} - J_\sigma) \left(\frac{z_i - z_j}{z_j} \right)^q. \end{aligned} \tag{4.22}$$

It is useful to consider the terms with $q = 0, 1$ and $q \geq 2$, separately. For the terms with $q = 0, 1$ two sets of site indices $\{\alpha\}$ and $\{\beta\}$ are introduced. The set $\{\alpha\}$ ($\{\beta\}$) includes all the locations of particles with the spin $\sigma(\sigma')$. Using the following identity

$$\left(\frac{z_k - z_i}{z_k - z_j} \right)^{\delta_{\sigma'\sigma_k} - \delta_{\sigma\sigma_k}} \equiv 1 - \delta_{\sigma'\sigma_k} \frac{z_i - z_j}{z_k - z_j} + \delta_{\sigma\sigma_k} \frac{z_i - z_j}{z_k - z_i} \quad \text{for } \sigma \neq \sigma', \tag{4.23}$$

the product in Eq. (4.18) may be expanded and the resulting terms in $P_{\{M\}}$ can be simplified using the following two theorems.

Theorem 4.1: *Let $\{\alpha\}$ and $\{\beta\}$ be two sets of distinct integers between 1 and N, and let $0 \leq q \leq 1$ and $\Delta = (1 - \delta_{\sigma,\sigma'})\delta_{\sigma\sigma_i}\delta_{\sigma\sigma_{\alpha_1}} \cdots \delta_{\sigma\sigma_{\alpha_n}}\delta_{\sigma'\sigma_j}\delta_{\sigma'\sigma_{\beta_1}} \cdots \delta_{\sigma'\sigma_{\beta_m}}$. Then,*

$$\begin{aligned} & \sum_{n=1}^{M_\sigma - 1} \sum_{m=1-q}^{M_{\sigma'} - 1} \sum_{i \neq j} \sum_{\{\alpha\},\{\beta\}} \frac{(-1)^m}{n!\,m!} \frac{z_i z_j^{1-q}(z_i - z_j)^{n+m-2+q}}{(z_{\alpha_1} - z_i) \cdots (z_{\alpha_n} - z_i)(z_{\beta_1} - z_j) \cdots (z_{\beta_m} - z_j)} \Delta \\ & = \begin{cases} -\sum_{k=1}^{\min(M_\sigma, M_{\sigma'})}(M_\sigma - k)(M_{\sigma'} - k) & \text{for } q = 0, \\ -\sum_{k=1}^{\min(M_\sigma, M_{\sigma'})}(M_\sigma - k) & \text{for } q = 1. \end{cases} \end{aligned}$$

Theorem 4.2: *For $q \geq 2$ the following identity will hold for $M_\sigma \geq M_{\sigma'}$,*

$$\sum_{i \neq j} \frac{z_i}{z_j} \left(\frac{z_i - z_j}{z_j} \right)^{q-2} (1 - \delta_{\sigma\sigma'})\delta_{\sigma\sigma_i}\delta_{\sigma'\sigma_j} \prod_{k \neq ij} \left(\frac{z_k - z_i}{z_k - z_j} \right)^{\delta_{\sigma'\sigma_k} - \delta_{\sigma\sigma_k}}$$

$$= \begin{cases} M_{\sigma'} & \text{for } q = 2 \\ 0 & \text{for } 2 < q \leq M_\sigma - M_{\sigma'} + 1. \end{cases}$$

Proofs of these theorems are given in Appendices 4.A and 4.B.

After some further reorganizing one may express the spin exchange operation $P_{\{M\}}$ as follow

$$P_{\{M\}} = W_P + U_P, \tag{4.24}$$

where

$$
\begin{aligned}
W_P &= -\tfrac{1}{3}\sum_\sigma M_\sigma(M_\sigma - 1)(M_\sigma - 2) - \tfrac{1}{3}\sum_{\sigma<\sigma'} M_{\sigma'}(M_{\sigma'} - 1)(3M_\sigma - M_{\sigma'} - 1), \\
&\quad - \sum_{\sigma<\sigma'} M_{\sigma'}(M_\sigma - M_{\sigma'})(J_{\sigma'} - J_\sigma) + \sum_{\sigma<\sigma'} M_{\sigma'}(J_{\sigma'} - J_\sigma)^2, \tag{4.25}
\end{aligned}
$$

$$
\begin{aligned}
U_P &= \sum_{i\neq j} \frac{z_i z_j}{(z_i - z_j)^2}\left\{1 - 2\delta_{\sigma_i\sigma_j}\right\} + \sum_{i\neq j} \frac{z_i J_{\sigma_j} - z_i J_{\sigma_i}}{z_i - z_j} \\
&\quad - 2\sum_{(i\neq j\neq k)} \frac{z_i z_j}{(z_i - z_j)(z_i - z_k)}\delta_{\sigma_i\sigma_k}. \tag{4.26}
\end{aligned}
$$

Because of symmetry I can choose $M_1 \geq M_2 \geq \cdots \geq M_n$ and $0 \leq J_{\sigma'} - J_\sigma \leq M_\sigma - M_{\sigma'} + 1$ for $\sigma' > \sigma$ without loss of generality. And, one can easily check that U_K given by Eq. (4.14), U_P above and the potential-energy term combine to give local variable independent terms.

Before I give the full expressions for the eigen-energy, it might be useful to discuss allowed values of the integer currents. Theorem 4.1 gives a simple rule for selecting allowed currents. If I choose $-1 \leq J_{\sigma'} - J_\sigma \leq M_\sigma - M_{\sigma'} + 1$ for $M_\sigma \geq M_{\sigma'}$, then $P_q = 0$ for $q > 2$. Otherwise, the energy will not in general be independent of the local variables and the corresponding wavefunction will not be an eigenfunction. Pictorial illustration of the allowed currents is shown in Fig. 4.2. Each row of M_ν boxes represents M_ν "particles" of the same spin. A single box gives a unit of current. Figure 4.2(a) is the reference state with no current. If the current is $j > 0$ (< 0) for a row, the row is moved j boxes to the right (left) as shown in Fig. 4.2(b). In order for a state to be an eigen-state, the following condition must hold true: *For a given pair of rows of boxes, all except the first and the last boxes in the row with smaller number of boxes must be positioned within the other row.* Figure 4.2(c), for example, cannot be an eigen-state because the last two boxes in the second row are not within the first row.

Finally, the eigen-energies of the $SU(n)$ Calogero-Sutherland model can be given as follow

$$E = \frac{2\hbar^2\pi^2}{mL^2}[E_1 + E_2 + E(x)], \tag{4.27}$$

where E_1 and E_2 are energies due to one- and two-spin interactions, respectively, and are given by

$$
\begin{aligned}
E_1 &= \tfrac{\lambda^2}{12}N(N^2 - 1) + \tfrac{1}{2}\lambda N\sum_\sigma M_\sigma(M_\sigma - 1) - \tfrac{1}{6}(\lambda - 1)\sum_\sigma M_\sigma(M_\sigma - 1)(2M_\sigma - 1) \\
&\quad + \sum_\sigma J_\sigma M_\sigma(M_\sigma + J_\sigma - 1), \tag{4.28}
\end{aligned}
$$

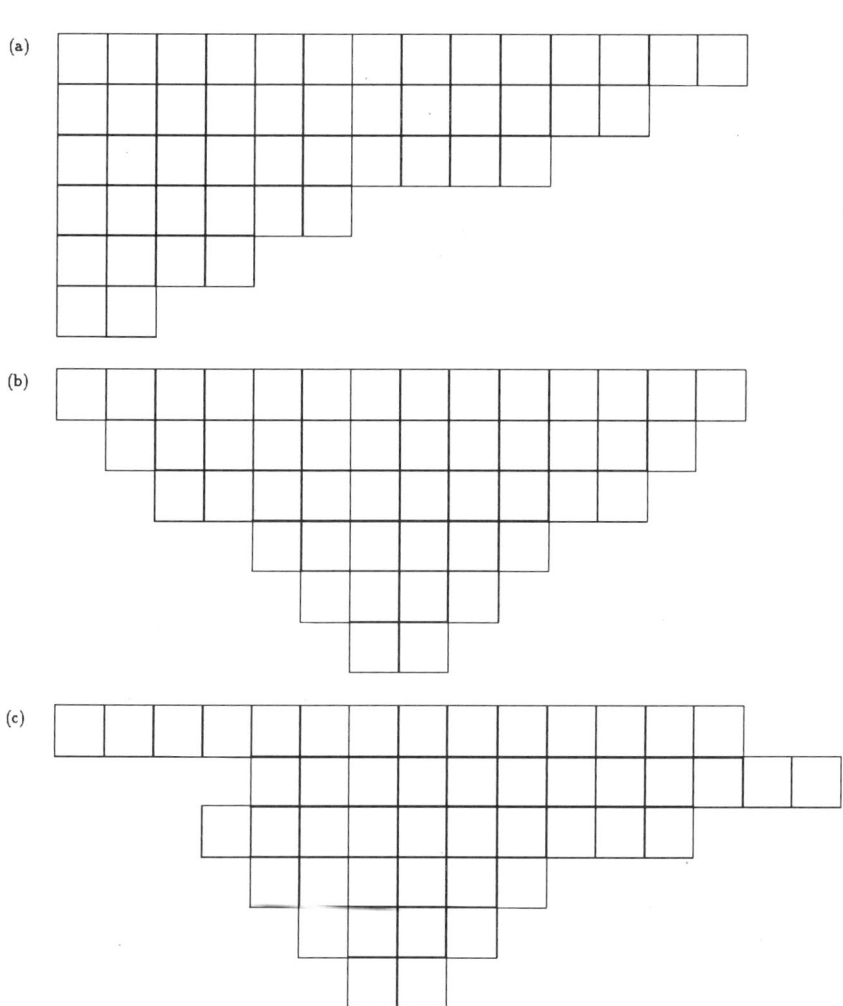

Fig. 4.2. Illustration of allowed values of the currents for the sector $\{14, 12, 10, 6, 4, 2\}$. (a) Allowed. All currents are zero. (b) Allowed. Ground state for the sector. (c) Not allowed.

$$E_2 = -\tfrac{1}{3}\lambda \sum_{\sigma<\sigma'} M_{\sigma'}(M_{\sigma'}-1)(3M_{\sigma}-M_{\sigma'}-1) - \lambda \sum_{\sigma<\sigma'} M_{\sigma'}(M_{\sigma}-M_{\sigma'})(J_{\sigma'}-J_{\sigma})$$
$$+ \lambda \sum_{\sigma<\sigma'} M_{\sigma'}(J_{\sigma'}-J_{\sigma})^2. \tag{4.29}$$

The statistical energy $E(x)$ is given by Eq. (4.15). And, the $SU(n)$ ground state is given by the following two conditions for all spin indices σ:

$$M_{\sigma} = M, \tag{4.30}$$
$$J_{\sigma} = -[J(x)], \tag{4.31}$$

where $N = nM$, $[y]$ is equal to the closest integer of y and

$$J(x) = \frac{x(N-1)+M-1}{2}. \tag{4.32}$$

Whenever $J(x)$ itself is an integer the total ground state energy is equal to the energy at $x = 0$. For a finite system, therefore, one might expect the ground state energy to be periodic with respect to x. In fact difference between the energies at arbitrary x and at $x = 0$ for M odd in units of $2\hbar^2\pi^2/mL^2$ is given by

$$\Delta E(x) = \frac{x^2}{4}N(N-1)^2 + N\left[\frac{x(N-1)}{2}\right]^2 - xN(N-1)\left[\frac{x(N-1)}{2}\right]. \tag{4.33}$$

Whenever the expression in the bracket above becomes an integer $\Delta E = 0$ and, thus, ΔE is periodic with period $2/(N-1)$. For $N = 21$, $M = 7$, and $n = 3$ the energy difference is plotted against the statistical parameter x in Fig. 4.3. For N large, however, this difference can be ignored since $\Delta E(x)$ increases at most linearly with N but the total energy increases as N^3.

4.3 $SU(n)$ t-J Model

In Ref. [40] the Calogero-Sutherland model that describes a system of spinless particles with ISE in continuum space was extended to the lattice. Using similar procedures the $SU(n)$ model described and solved in the previous section can be generalized to the lattice. The following hamiltonian is proposed in [31]

$$H = \mathcal{P}\left(\sum_{j=1}^{N}\sum_{n=1}^{N_a-1}\frac{t_j^n}{d(n)^2} + \sum_{i>j}\frac{l(l+P_{ij}^{\sigma})n_in_j}{d(x_i-x_j)^2}\right)\mathcal{P}, \tag{4.34}$$

where $d(n) = (N_a/\pi)|\sin(n)|$, l is a positive integer, N_a is the total number of sites and N the total number of particles. The operator t_j^n hops a particle at the jth site to the $(j+n)$th (mod N_a) site and n_j is the occupation number at jth site. P_{ij}^{σ} is, as before, the spin exchange operator. $\mathcal{P}$ is the projection operator that insures the absence of multiply occupied sites by projecting them out of the Hilbert space.

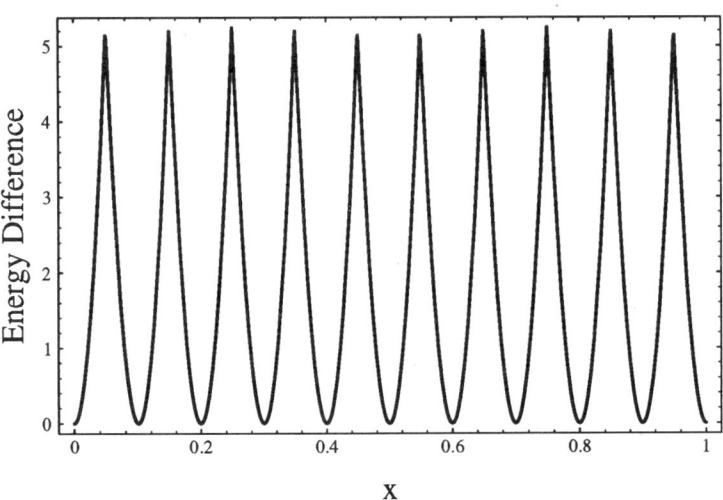

Fig. 4.3. The energy difference ΔE in units of $2\hbar^2\pi^2/mL^2$ vs. the statistical parameter x for $N = 21$, $M = 7$, and $n = 3$. Note the period $1/10$.

In Ref. [31] the following eigen-states for the model hamiltonian are proposed

$$|\Psi_{\{M\}}\rangle = \sum_{\{z\sigma\}} \phi(\{z\sigma\})|\{z\sigma\}\rangle. \tag{4.35}$$

Since the total number of each spin is a good quantum number, one can classify the eigen-states into sectors labeled by $\{M\} \equiv (M_1, M_2, \ldots, M_n)$, where M_ν is the number of νth spin and $\sum_{\nu=1}^{n} M_\nu = N$ and also represent the particle configuration as $\{z\sigma\} \equiv (z_1\sigma_1, \ldots, z_i\sigma_i, \ldots, z_j\sigma_j, \ldots, z_N\sigma_N)$ where z_i and σ_i are the location and spin of the ith particle. The sum in Eq. (4.35) is over $N_a!/(M_h!M_1! \cdots M_n!)$ distinct spin configurations of the sector $\{M\}$ where N_a and M_h are the total number of sites and the number of holes, respectively. The function ϕ in Eq. (4.35) is given by the following Jastrow-product:

$$\phi(\{z\}|\{\sigma\}) = \prod_{i<j}(z_i - z_j)^{l+\delta_{\sigma_i\sigma_j}} e^{i\frac{\pi}{2}\mathrm{sgn}(\sigma_i-\sigma_j)} \prod_k z_k^{J_{\sigma_k}}. \tag{4.36}$$

Considering the symmetry of the wavefunction one may take odd (even) l to correspond to the bosonic (fermionic) case.

As in the continuum case the total hamiltonian is divided into the kinetic, potential, and spin exchange parts—H_L^0, H_L^1 and H_L^3, respectively. The actions of H_L^1 and H_L^3 are same as the corresponding operators in the continuum case. For H_L^0 one has the following relations:

$$\langle H_L^0 \rangle \equiv \frac{\langle z_1\sigma_1, \ldots, z_N\sigma_N | H_L^0 | \Psi_{\{M\}} \rangle}{\langle z_1\sigma_1, \ldots, z_N\sigma_N | \Psi_{\{M\}} \rangle}$$

$$= 4\sum_{i=1}^{N} \sum_{n=1}^{N_a-1} z^{nJ_{\sigma_i}}(1-z^n)^{-1}(1-z^{-n})^{-1} \prod_{j(\neq i)} \left(\frac{z_i z^n - z_j}{z_i - z_j} \right)^{l+\delta_{\sigma_i\sigma_j}} \quad (4.37)$$

where $|z_1\sigma_1, \ldots, z_N\sigma_N\rangle$ is one of the basis states of the sector $|\Psi_{\{M\}}\rangle$. In order to evaluate $\langle H_L^0 \rangle$ the following theorem is useful (a slightly different version of this theorem can be found in Ref. [40]).

Theorem 4.3: *Let J and p be non-negative integers and $z = \exp(2\pi i/N_a)$. Then, the sum*

$$\sum_{n=1}^{N_a-1} z^{nJ}(1-z^{-n})^{-1}(1-z^n)^{p-1} \quad (4.38)$$

is equal to the following in various different cases

$$\frac{N_a^2 - 1}{12} - \frac{J(N_a - J)}{2} \quad for \quad p = 0,$$

$$-J + \frac{N_a - 1}{2} \quad for \quad p = 1,$$

$$+1 \quad for \quad p = 2,\ 0 \le J \le N_a - 2$$

$$-(N_a - 1) \quad for \quad p = 2,\ J = N_a - 1$$

$$\sum_{m>0}(-1)^{mN_a - J}\binom{p-2}{mN_a - J - 1}N_a \quad for \quad p \ge 3,$$

where m is a positive integer and the restriction on the current is $0 \le J \le N_a - 1$ unless specified otherwise.

The proof of this theorem is rather straightforward and, thus, will not be given here.

If l in Eq. (4.34) is a positive integer one may multiply out the product in Eq. (4.37). The resulting expression will be a sum of terms containing a factor $(1 - z^n)^p$. Since the maximum value of p is $l(N-1) + M_\sigma - 1$, one may ignore terms with $p \ge 3$ if $0 \le J_\sigma \le N_a - l(N-1) - M_\sigma + 1$. I can combine the remaining terms with those from H_L^1 and H_L^2 and obtain the following eigen-energies for the lattice model

$$E^L = E_1^L + E_2^L, \quad (4.39)$$

where

$$E_1^L = \tfrac{1}{6}N(N_a^2 - 1) + \frac{l^2}{3}N(N-1)(N-2) - \frac{l}{2}(N_a - l)N(N-1)$$

$$- \left(\frac{N_a - 1}{2} - l(N-1) \right)\sum_\sigma M_\sigma(M_\sigma - 1) + \tfrac{1}{3}(1-l)\sum_\sigma M_\sigma(M_\sigma - 1)(M_\sigma - 2)$$

$$- \sum_\sigma M_\sigma J_\sigma(N_a - lN - M_\sigma - J_\sigma + l + 1), \quad (4.40)$$

$$E_2^L = E_2, \quad (4.41)$$

where E_2 is given by Eq. (4.29). It is emphasized at this point that the above expressions for the eigen-energies for the lattice model are valid only for $lN + M_\sigma \leq N_a + l + 1$.

For $l = 1$ (bosonic supersymmetric case) one can find another expression for the energy for $N + M_\sigma \geq N_a + 2$ by first rewriting the kinetic-energy term as

$$\langle H^0 \rangle = 2 \sum_\sigma \sum_{i \neq j} (1 - \delta_{\sigma\sigma_h}) \delta_{\sigma\sigma_i} \delta_{\sigma_h\sigma_j} \left(\frac{z_i}{z_j} \right)^{J_\sigma + 1} \frac{z_i z_j}{(z_i - z_j)^2} \prod_{k \neq ij} \left(\frac{z_k - z_i}{z_k - z_j} \right)^{\delta_{\sigma_h\sigma_k} - \delta_{\sigma\sigma_k}},$$
(4.42)

where σ_h represents the hole. The above expression is almost identical to the expression for the spin exchange operator and can easily be evaluated using Theorems 4.1 and 4.2. The energy is then given by

$$E(l = 1) = \frac{\pi^2}{2N_a^2} [E_k(l = 1) + W_P],$$
(4.43)

where

$$\begin{aligned}
E_k(l = 1) &= -\tfrac{1}{6} M_h(N_a^2 - 1) - \tfrac{1}{3} M_h(M_h - 1)(M_h - 2) + M_h \sum_\sigma J_\sigma(J_\sigma - M_\sigma - 1) \\
&+ \tfrac{1}{2} N_a - 1 \left(\sum_\sigma M_\sigma(M_\sigma - 1) + M_h(M_h - 1) \right) + M_h N \\
&+ \sum_\sigma (J_\sigma - 1)(2M_\sigma M_h - M_h(M_h + 1)) \\
&- \tfrac{1}{3} M_h(M_h - 1)[3N - n(M_h + 1)],
\end{aligned}$$
(4.44)

where the currents are restricted to $-(M_\sigma - M_h) \leq J_\sigma \leq 0$. W_P in Eq. (4.43) is given by Eq. (4.25).

For $l = 0$ the wavefunction in Eq. (4.35) is that of Gutzwiller projected free fermions and no longer an eigen-state of the hamiltonian in Eq. (4.34). However, the wavefunction is an eigen-state of the following hamiltonian:

$$H = P \left(-\sum_{j,n} \frac{t_j^n}{d(n)^2} + \sum_{i>j} \frac{(P_{ij}^\sigma - 1) n_i n_j}{d(x_i - x_j)^2} \right) P.$$
(4.45)

This hamiltonian is a $SU(n)$ generalization of the $SU(2)$ fermionic supersymmetric t-J model [70, 31]. If a similar technique used for the $l = 1$ case is employed one may find the eigen-energy given by Eq. (4.43) with $E_k(l = 1)$ replaced with $E_k(l = 0)$, which is given by

$$E_k(l = 0) = -\frac{N(N_a^2 - 1)}{6} + \tfrac{1}{2}(N_a - 1) \sum_\sigma M_\sigma(M_\sigma - 1) + M_h \sum_\sigma J_\sigma(M_\sigma + J_\sigma - 1).$$
(4.46)

The currents are integers restricted to $-M_\sigma \leq J_\sigma \leq 0$. The ground-state energy is obtained if $J_\sigma = -(M_\sigma - 1)/2$ and M_σ odd. If M_σ are not all odd, the ground state is degenerate.

4.4 $SU(n)$ **Spin Chain**

The spin exchange hamiltonian in the model discussed in the previous section corresponds to the full hamiltonian for the $SU(n)$ spin chain. This is equivalent to taking the "half-filling" limit $(N \rightarrow N_a)$ of the lattice t-J model. Since every site is occupied with a spin, U_P in Eq. (4.26) can now be summed and is equal to the following

$$- N_a(N_a^2 - 1)/12 + \{(N_a - 1)/2\} \sum_\sigma M_\sigma(M_\sigma - 1). \tag{4.47}$$

Hence, U_P is now local variable independent, and the eigen-energy is given by Eq. (4.24).

One can also consider the $SU(n)$ spin chain as $SU(n-1)$ supersymmetric bosonic t-J model by regarding one of the species as hole and the rest as particles as discussed in Chapter 2 for $SU(2)$ case. Details of the energy spectra and the quantum symmetry are discussed in Chapter 5. The spin chain can also be generalized to graded $SU(n|m)$ supersymmetric t-J model where there are n species of bosons and m species of fermions. This generalization especially the more familiar $SU(1|2)$ case is discussed in great detail in Chapter 6.

4.5 **Twisted Boundary Condition**

So far only the periodic boundary condition have been considered. In this section let us expand our horizon by considering a more general twisted boundary condition which arises when the magnetic field pierces perpendicular to the plane of 1D ring. That is when a particle goes around the entire system once it picks up a net phase $\exp(i\phi)$ and if it goes around n times it picks up $\exp(i\phi n)$ extra phase. One could therefore generalize the sum given in Eq. (4.2) to the following

$$\sum_{n=-\infty}^{\infty} \frac{e^{i\phi n}}{(x + nL)^2} \tag{4.48}$$

If the phase angle ϕ is a rational fraction of 2π the sum above can be explicitly performed. Let $\phi = 2\pi p/q$ and $n = jq + k$ where p and q are mutual primes and j and k are integers such that $-\infty < j < +\infty$ and $0 \le k \le q-1$. The sum above then becomes

$$\sum_{k=0}^{q-1} \sum_{j=-\infty}^{\infty} \frac{1}{((x + kL) + (qL)j)^2}. \tag{4.49}$$

Hence, by using Eq. (4.2) one easily obtains the following final expression

$$\sum_{k=1}^{q-1} \frac{e^{i2\pi pk/q}}{\left\{ \frac{qL}{\pi} \sin\left(\frac{\pi(x+kL)}{qL} \right) \right\}^2}. \tag{4.50}$$

The effective length of the chain is thus qL. When $p/q = 1/2$ (i.e. anti-periodic boundary condition) the corresponding particle interaction term becomes

$$\frac{1}{(d'(x))^2} = \frac{\cos(\pi x/L)}{(d(x))^2}, \tag{4.51}$$

where $d(x)$ is the chord distance as before. The t-J model hamiltonian can thus be generalized to

$$H = \mathcal{P}\left(\sum_{j=1}^{N}\sum_{n=1}^{N_a-1}\frac{t_j^n}{d'(n)^2} + \sum_{i>j}\frac{l(l+P_{ij}^{\sigma})n_i n_j}{d(x_i-x_j)^2}\right)\mathcal{P}, \tag{4.52}$$

4.6 Harmonic Fluid Description

Let us now turn our attention to the low-energy excitation spectrum of our model in the thermodynamic limit. The low-energy excitation spectrum of a one-dimensional quantum fluid may be described by a sum of two independent harmonic fluid hamiltonians, one for the charge and the other for the spin. This is a slight generalization of the spinless particle system in Ref. [38]. In general the effective hamiltonian can be written as follow

$$H = \sum_{\sigma,\sigma'}\int dx\left\{A_{\sigma\sigma'}\Pi_\sigma(x)\Pi_{\sigma'}(x) + B_{\sigma\sigma'}\nabla\phi_\sigma(x)\nabla\phi_{\sigma'}(x)\right\}, \tag{4.53}$$

where $\Pi_\sigma(x)$ is the local-density-fluctuation field as in Ref. [38] and $\phi_\sigma(x)$ is the canonical conjugate field to $\Pi(x)$, i.e. obey the following commutation relation

$$[\phi_\sigma(x),\Pi_{\sigma'}(x')] = i\delta_{\sigma\sigma'}\delta(x-x'). \tag{4.54}$$

Because of the $SU(n)$ symmetry one can write the couplings A and B as

$$A_{\sigma\sigma'} = a_1 + a_0\delta_{\sigma\sigma'}, \tag{4.55}$$
$$B_{\sigma\sigma'} = b_1 + b_0\delta_{\sigma\sigma'}, \tag{4.56}$$

and can express the hamiltonian in terms of the Fourier-transformed fields, $\Pi_{\sigma k}$ and $\phi_{\sigma k}$, as

$$H = \sum_{k}\sum_{\sigma,\sigma'}A_{\sigma\sigma'}\Pi_{\sigma k}\Pi_{\sigma'-k} + B_{\sigma,\sigma'}k^2\phi_{\sigma k}\phi_{\sigma'-k}. \tag{4.57}$$

One may construct normal mode fields as $\Pi_k^\nu = \sum_\sigma a_\sigma^\nu\Pi_{\sigma k}$ and $\phi_k^\nu = \sum_\sigma b_\sigma^\nu\phi_{\sigma k}$. If one chooses $\sum_\sigma b_\sigma^\nu a_\sigma^{\nu'} = \delta_{\nu\nu'}$, then $[\phi_k^\nu,\Pi_{-k'}^{\nu'}] = i\delta_{kk'}\delta_{\nu\nu'}$. One then has the following equations of motion

$$[H,[H,\Pi_k^\nu]] = -(v^\nu k)^2\Pi_k^\nu, \tag{4.58}$$
$$[H,[H,\phi_k^\nu]] = -(v^\nu k)^2\phi_k^\nu. \tag{4.59}$$

The Π field satisfies the following equation,

$$(a_0 b_0 - (v^\nu)^2) a_\sigma^\nu + (a_1 b_0 + a_0 b_1 + n a_1 b_1) \sum_\beta a_\beta^\nu = 0. \tag{4.60}$$

The same equation with a_σ^ν replaced with b_σ^ν holds for the ϕ field. There are only two possible values for $(v^\nu)^2$: $a_0 b_0$ and $(a_0 + n a_1)(b_0 + n b_1)$. The first value corresponds to the case $\sum_\beta a_\beta^\nu = 0$ and the second $\sum_\beta a_\beta^\nu \neq 0$. Hence, the first would be the spin velocity and the second the charge velocity. The hamiltonian can now be written as

$$H = \sum_k \left[(a_0 + n a_1) \Pi_k^c \Pi_{-k}^c + (b_0 + n b_1) k^2 \phi_k^c \phi_{-k}^c + \sum_{s=1}^{n-1} \left\{ a_0 \Pi_k^s \Pi_{-k}^s + b_0 k^2 \phi_k^s \phi_{-k}^s \right\} \right] \tag{4.61}$$

There are one charge mode and $n-1$ spin modes; and because of the $SU(n)$ symmetry all the spin modes have the same velocity. Let us define the following velocities associated with Π and ϕ fields for the charge (denoted as c) and spin (denoted as s) modes

$$v_N^c = a_0 + n a_1, \tag{4.62}$$
$$v_J^c = b_0 + n b_1, \tag{4.63}$$
$$v_N^s = a_0, \tag{4.64}$$
$$v_J^s = b_0. \tag{4.65}$$

The charge and spin velocities then satisfy the following harmonic fluid relations

$$v_c = (v_N^c v_J^c)^{1/2}, \tag{4.66}$$
$$v_s = (v_N^s v_J^s)^{1/2}. \tag{4.67}$$

Since the eigen-states for our model correspond to the low-energy excitations of the model, we can easily obtain the harmonic fluid parameters by examining the energy expanded about the ground state. By Taylor expansion we obtain the energy for the continuum model as follow

$$E = e_0 N + \mu N + \frac{\pi \hbar \rho_0}{m} \frac{\pi \hbar}{L} \sum_{\sigma \sigma'} \tfrac{1}{2} A_{\sigma \sigma'} \Delta M_\sigma \Delta M_{\sigma'} + 2 B_{\sigma \sigma'} \Delta J_\sigma \Delta J_{\sigma'} \tag{4.68}$$

where

$$e_0 = \frac{(\pi \hbar \rho_0)^2}{6m} \left(\lambda + \frac{1}{n} \right)^2, \tag{4.69}$$
$$\mu = \frac{(\pi \hbar \rho_0)^2}{2m} \left(\lambda + \frac{1}{n} \right)^2, \tag{4.70}$$
$$A_{\sigma \sigma'} = \lambda \left(\lambda + \frac{1}{n} \right) + \left(\lambda + \frac{1}{n} \right) \delta_{\sigma \sigma'}, \tag{4.71}$$
$$B_{\sigma \sigma'} = -\frac{\lambda}{n} + \left(\lambda + \frac{1}{n} \right) \delta_{\sigma \sigma'}, \tag{4.72}$$

where $\rho_0 = N/L$. One can thus easily read off the following velocities,

$$v_s = v_N^s = v_J^s = \frac{\pi \hbar \rho_0}{m}\left(\lambda + \frac{1}{n}\right), \tag{4.73}$$

$$v_N^c = \frac{\pi \hbar \rho_0}{m} n \left(\lambda + \frac{1}{n}\right)^2, \tag{4.74}$$

$$v_J^c = \frac{\pi \hbar \rho_0}{m} \frac{1}{n}, \tag{4.75}$$

$$v_c = v_s. \tag{4.76}$$

The charge and spin velocities are the same for all λ and n. As expected from the singlet nature of the ground state, the ratio v_J^s/v_N^s does not get renormalized. The coefficient v_J^c is independent of the interaction term, as a consequence of Galilean invariance, but v_N^c gets renormalized due to nonlinear interaction terms not included in the linear form of the harmonic hamiltonian. The renormalization coupling constant for the charge is $\exp(-2\varphi) \to (v_N^c/v_J^c)^{1/2} = n\lambda + 1$ in the limit of long wavelength (i.e., $k \to 0$). Because of the scale invariance of the model the dimension-less coupling constant φ is independent of the particle density.

The compressibility per particle is $(\pi^2 \hbar^2 \rho_0^2/m)(\lambda + 1/n)^2$. One can also find the chemical potential and the ground-state energy to be $m^* v_F^2/2$ and $(N/3)(m^* v_F^2/2)$, respectively. The chemical potential (or the Fermi energy at zero temperature) and the ground-state energy of this one-dimensional system are that of free fermions with the renormalized mass per particle $m^* = m/(n\lambda + 1)^2$. The Fermi velocity is given by $v_F = \pi \hbar \rho_0/(nm^*)$.

The low-energy excitation of the lattice version is given by Eq. (4.68) as well if one sets $m = \hbar = 1$ and replaces L, e_0, and μ with N_a, $-\pi^2/3 + 2e_0$, and $-\pi^2/3 + 2\mu$, respectively. The crucial difference is that this expression for the lattice model is true only for $\rho_0(= N/N_a) \le \rho_0^{\max}[= (l + 1/n)^{-1}]$. Hence, the charge velocity is linear in ρ_0 only up to $\rho_0^{\max}$. We expect v_c to vanish as $\rho_0 \to 1$ due to the metal-insulator transition at the density. The charge velocity v_c, therefore, should exhibit non-analyticity at $\rho_0^{\max}$. This behavior is attributed to the long-range interaction in our model. In Ref. [45] this type of non-analyticity in the spinon velocity was observed for the $SU(2)$ spin chain model. In the nearest-neighbor interaction models the sharp change in the charge velocity is smoothed out.

For $l = 1$ the energy for $\rho_0 \ge (1 + 1/n)$ [1] is explicitly found and the harmonic fluid hamiltonian is given by Eq. (4.68) with the following couplings

$$A_{\sigma\sigma'} = (1 - \rho_0)(n + 2) - \rho_0/n + \delta_{\sigma\sigma'}, \tag{4.77}$$

$$B_{\sigma\sigma'} = -\rho_0/n + \delta_{\sigma\sigma'}. \tag{4.78}$$

Therefore, $v_c = \pi(1 - \rho_0)(n - 1)$ and $v_s = \pi$. The spin velocity is independent of the density in this region. Similarly, for $l = 0$ we have $A_{\sigma\sigma'} = B_{\sigma\sigma'} = -\rho_0/n + \delta_{\sigma\sigma'}$ for $0 \le \rho_0 \le 1$. Thus, $v_c = \pi(1 - \rho_0)$ and $v_s = \pi$. It is somewhat unexpected that the charge and spin velocities are independent of the number of spin species.

This feature must be due to the elementary excitations forming free gas. Figure 4.4 shows the charge and spin velocities for the lattice models with various interaction parameters.

4.7   Conclusion

In this chapter the Calogero-Sutherland model of spinless particles interacting with inverse-square exchange is shown to be generalizable to the multicomponent system of particles. It is further generalized from the continuum model to the lattice for integer values of the interaction parameter λ. A class of eigen-states which correspond to the ground state and low-energy excitations for these models is constructed and the corresponding eigen-energies are also calculated. These results are checked by the exact numerical diagonalization of the lattice hamiltonian for small systems. The systematic construction of the other eigen-states, thermodynamics, and the general n-point correlation functions for these models are yet to be found. The hamiltonian of the type given by Eq. (4.1) also arises in a matrix model as a representation of one-dimensional open string theory.

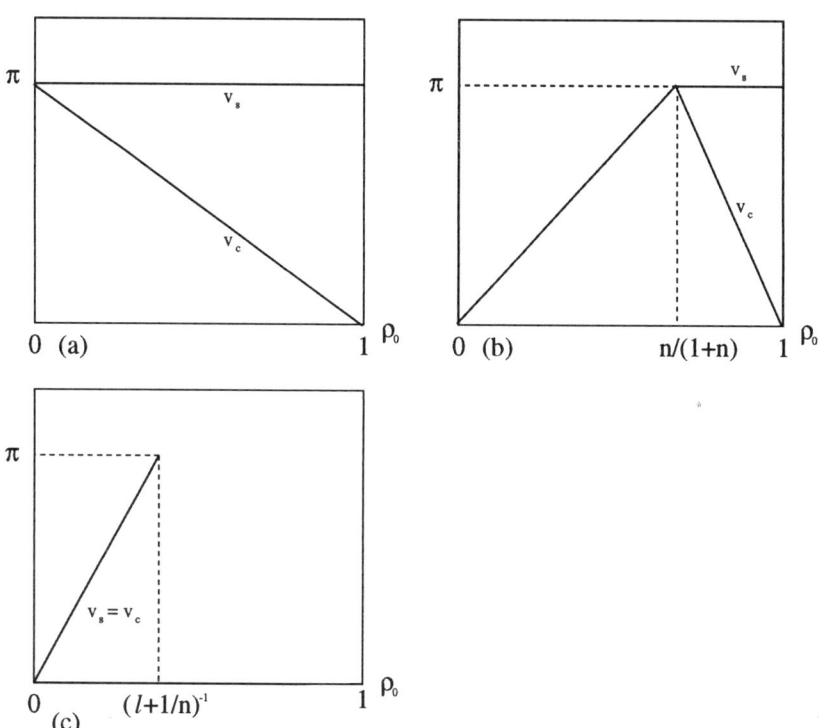

Fig. 4.4. Charge and spin velocities of the lattice models (a) Fermionic supersymmetric t-J model. The velocities are independent of the number of spin species. (b) Bosonic supersymmetric t-J model. Up to the density $n/(1+n)$, v_c and v_s are equal. (c) Bosonic and fermionic t-J models with $l \geq 2$. The charge and spin velocities are the same up to the density $(l+1/n)^{-1}$. Beyond the density the ground state is no longer the Jastrow-product type.

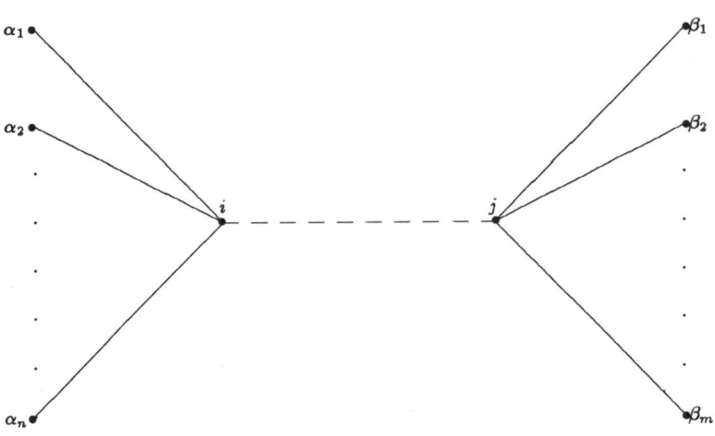

Fig. 4.5. Labeled diagram for n σ-spins and m σ'-spins.

4.A   Proof of Theorem 3.1

We first discuss a diagrammatic way of writing the exchange operation. A labeled diagram is shown in Fig. 4.5. The amplitude of the labeled diagram shown in Fig. 4.5 can be evaluated using the following rules: (i) the dashed line connecting the indices i and j gets a factor $z_i z_j (z_i - z_j)^{-2} (1 - \delta_{\sigma,\sigma'}) \delta_{\sigma\sigma_i} \delta_{\sigma\sigma_j}$; (ii) each solid line connecting the indices i and α_k gets a factor $(z_i - z_j)(z_{\alpha_k} - z_i)^{-1} \delta_{\sigma\sigma_{\alpha_k}}$; (iii) each solid line connecting the indices j and β_k gets a factor $-(z_i - z_j)(z_{\beta_k} - z_j)^{-1} \delta_{\sigma\sigma_{\beta_k}}$. Weight of the dashed line vanishes whenever the spins at the sites i and j are the same. On the other hand, weights of the solid lines that are connected to $i(j)$ vanish whenever the spins at the sites $\{\alpha\}$ ($\{\beta\}$) and $i(j)$ are different. Hence, we may think of the diagram as a two-spin interaction diagram. Since the hamiltonian has only the two-spin exchange operator P_{ij}, three-spin and higher number of spin interactions are missing.

The amplitude of the labeled diagram is undesirable since it depends on the local variables. A more desirable, local-variable-independent diagram (unlabeled diagram) can be obtained by summing $(n + 1)(m + 1)$ cyclically permuted labeled diagrams. This fact will be proved in this appendix.

We note that the indices $\{i, \alpha_1, \ldots, \alpha_n\}$ and $\{j, \beta_1, \ldots, \beta_m\}$ are dummy indices. Consequently, the sum will be invariant under any permutation of the indices. We further observe that Δ is invariant under any permutation of the indices except when there is interchange between the two sets $\{i, \alpha_1, \ldots, \alpha_n\}$ and $\{j, \beta_1, \ldots, \beta_m\}$. On the other hand, the prefactor to Δ remains invariant with respect to the permutations of indices $\{\alpha\}$ and $\{\beta\}$. It is convenient to consider only the cyclic permutations which give $(n+1)(m+1)$ distinct prefactors. We then sum the $(n+1)(m+1)$ factors as follow:

$$
\begin{aligned}
P_{nm}^{\sigma\sigma'} &= \sum_{i\neq j} \sum_{\{\alpha\},\{\beta\}} \frac{(-1)^m}{n!\,m!} \frac{z_i z_j^{1-q}(z_i - z_j)^{n+m-2+q}}{(z_i - z_{\alpha_1})\cdots(z_i - z_{\alpha_n})(z_j - z_{\beta_1})\cdots(z_j - z_{\beta_m})}\Delta \\
&= \sum_{i\neq j} \sum_{\{\alpha\},\{\beta\}} \sum_{k,l} \frac{(-1)^m}{(n+1)!\,(m+1)!} \frac{z_{\gamma_k^\alpha} z_{\gamma_l^\beta}^{1-q}(z_{\gamma_k^\alpha} - z_{\gamma_l^\beta})^{n+m-2+q}}{(z_{\gamma_k^\alpha} - z_{\alpha_1})\cdots(z_{\gamma_k^\alpha} - z_i)\cdots(z_{\gamma_k^\alpha} - z_{\alpha_n})} \\
&\quad\times \frac{1}{(z_{\gamma_l^\beta} - z_{\beta_1})\cdots(z_{\gamma_l^\beta} - z_j)\cdots(z_{\gamma_l^\beta} - z_{\beta_m})}\Delta,
\end{aligned}
\tag{4.A.79}
$$

where $\{\gamma^\alpha\} \equiv \{i, \alpha_1, \ldots, \alpha_n\}$ and $\{\gamma^\beta\} \equiv \{j, \beta_1, \ldots, \beta_m\}$.

We let

$$
D_\alpha = \prod_{k<k'} (z_{\gamma_k^\alpha} - z_{\gamma_{k'}^\alpha}),
\tag{4.A.80}
$$

$$
D_\beta = \prod_{l<l'} (z_{\gamma_l^\beta} - z_{\gamma_{l'}^\beta}).
\tag{4.A.81}
$$

Using the binomial theorem we expand the numerator in Eq. (4.A.79) and, then, multiply and divide the expression by $D_\alpha D_\beta$ and obtain the following:

$$
P_{nm}^{\sigma\sigma'} = \sum_{i\neq j} \sum_{\{\alpha\},\{\beta\}} \sum_{s=1}^{n+m-1+q} \frac{(-1)^{m-1+q-s}}{(n+1)!\,(m+1)!} \binom{n+m-2+q}{s-1} \frac{V_\alpha^s V_\beta^{n+m-s}}{D_\alpha D_\beta}\Delta,
\tag{4.A.82}
$$

where

$$
V_\alpha^s = \begin{vmatrix} 1 & 1 & \cdots & 1 \\ z_i & z_{\alpha_1} & \cdots & z_{\alpha_n} \\ \vdots & \vdots & & \vdots \\ z_i^{n-1} & z_{\alpha_1}^{n-1} & \cdots & z_{\alpha_n}^{n-1} \\ z_i^s & z_{\alpha_1}^s & \cdots & z_{\alpha_n}^s \end{vmatrix},
\tag{4.A.83}
$$

and V_β^k is defined the same way. We further note that $V_\alpha^s = 0$ if $0 \leq s \leq n-1$ and $V_\beta^k = 0$ if $0 \leq k \leq m-1$. Therefore, if $0 \leq q \leq 1$, the only nonzero contribution in (4.A.82) is the term with $s = n$. Since $V_\alpha^n = D_\alpha$ and $V_\beta^m = D_\beta$, we obtain

$$
\begin{aligned}
P_{nm}^{\sigma\sigma'} &= \frac{(-1)^{n+m-1+q}}{(n+1)!\,(m+1)!} \binom{n+m-2+q}{n-1} \sum_{i\neq j} \sum_{\{\alpha\},\{\beta\}} \Delta \\
&= (-1)^{n+m-1+q} \binom{n+m-2+q}{n-1} \binom{M_\sigma}{n+1} \binom{M_{\sigma'}}{m+1}.
\end{aligned}
\tag{4.A.84}
$$

Evaluating $\sum_{i\neq j} \sum_{\{\alpha\},\{\beta\}} \Delta$ in (4.A.84) is the same as calculating the total number of ways of putting $n+1$ out of M_σ blue balls in a box and $m+1$ out of $M_{\sigma'}$ red balls in another box. The sum, therefore, is equal to $M_\sigma(M_\sigma - 1)\cdots(M_\sigma - n)M_{\sigma'}(M_{\sigma'} - 1)\cdots(M_{\sigma'} - m)$.

It is straightforward to prove the following useful identities:

$$\binom{n+m-2+q}{n-1} = \sum_{s=0}\binom{n-1}{s}\binom{m-1+q}{s},\tag{4.A.85}$$

$$(-1)^s\binom{m-1+q}{s} = \sum_{k=0}^{s}(-1)^k\binom{m+q}{k},$$

$$= \sum_{k=0}^{s}(-1)^k(s+1-k)\binom{m+1+q}{k},\tag{4.A.86}$$

$$\sum_{m=1-q}^{M_\sigma-1}(-1)^m\binom{M_\sigma}{m+1}\binom{m+1}{k} = \left\{\begin{array}{ll} \left.\begin{array}{ll} -M_\sigma+1 & \text{if}\quad k=0 \\ -M_\sigma & \text{if}\quad k=1 \\ 0 & \text{if}\quad 2\le k< M_\sigma \end{array}\right\} & \text{for } q=0, \\ \left.\begin{array}{ll} 1 & \text{if}\quad k=0 \\ 0 & \text{if}\quad 1\le k< M_\sigma \end{array}\right\} & \text{for } q=1, \end{array}\right.\tag{4.A.87}$$

Using (4.A.85) we may sum the terms depending on n and m in $P_{nm}^{\sigma\sigma'}$ separately. Using (4.A.86) and (4.A.87) the following relation can easily be obtained:

$$P^{\sigma\sigma'} = \left\{\begin{array}{ll} -\sum_{k=1}^{\min(M_\sigma,M_{\sigma'})}(M_\sigma - k)(M_{\sigma'} - k) & \text{for } q=0, \\ -\sum_{k=1}^{\min(M_\sigma,M_{\sigma'})}(M_\sigma - k) & \text{for } q=1. \end{array}\right.\tag{4.A.88}$$

Q.E.D.

4.B   Proof of Theorem 3.2

We replace the product over k with two new sets of integer indices, $\{\alpha\}$ and $\{\beta\}$, and rewrite the expression in theorem 2 as,

$$Q = \sum_{i\ne j}\sum_{\{\alpha\},\{\beta\}}\frac{1}{(M_\sigma - 1)!\,(M_{\sigma'} - 1)!}\frac{z_i}{z_j}\left(\frac{z_i - z_j}{z_j}\right)^{q-2}\frac{z_{\alpha_1} - z_j}{z_{\alpha_1} - z_i}\cdots\frac{z_{\alpha_{M_\sigma-1}} - z_j}{z_{\alpha_{M_\sigma-1}} - z_i}$$

$$\times\left\{\sum_{r=0}^{M_{\sigma'}-1}(z_i - z_j)^r f_r(\{z_\beta\})\right\}\Delta.\tag{4.B.89}$$

Here, Δ is the same as in theorem 1. The factor in the curly brackets in (4.B.89) is obtained by writing the terms like $(z_{\beta_l} - z_i)/(z_{\beta_l} - z_j)$ as $\{1 - (z_i - z_j)/(z_{\beta_l} - z_j)\}$ and by multiplying out all the factors that depend on $\{\beta\}$. f_r in (4.B.89) is some function of $\{z_\beta\}$, however $f_0 = 1$.

As in Appendix 4.A, we rewrite the expression by summing over the factors obtained by the cyclic permutations of the indices $\{i,\alpha_1,\ldots,\alpha_{M_\sigma-1}\}$ and obtain

$$Q = \sum_{i\ne j}\sum_{\{\alpha\},\{\beta\}}\sum_{r=0}^{M_{\sigma'}-1}\frac{(-1)^{M_\sigma}}{M_\sigma!\,(M_{\sigma'} - 1)!}(z_i - z_j)(z_{\alpha_1} - z_j)\cdots(z_{\alpha_{M_\sigma-1}} - z_j)f_r\frac{W_\alpha^{M_\sigma}}{D_\alpha^{M_\sigma}},\tag{4.B.90}$$

where $W_\alpha^{M_\sigma}$ is given by the Vandemonde determinant whose last row is modified to $z_{\gamma_l}(z_{\gamma_l} - z_j)^{q+r-3}/z_j^{q-1}$, $\{\gamma_l\} \equiv \{i, \alpha_1, \ldots, \alpha_{M_\sigma-1}\}$. It is straightforward to show that

$$W_\alpha^{M_\sigma} = \begin{cases} \dfrac{(-1)^{M_\sigma} D_\alpha^{M_\sigma}}{(z_i-z_j)(z_{\alpha_1}-z_j)\cdots(z_{\alpha_{M_\sigma}-1}-z_j)} & \text{if } q = 2 \text{ and } r = 0 \\ 0 \text{ if } 0 \leq q + r - 3 \leq M_\sigma - 3 \end{cases}, \quad (4.B.91)$$

$$\sum_{i \neq j} \sum_{\{\alpha\},\{\beta\}} \Delta = M_\sigma! \, M_{\sigma'}!. \quad (4.B.92)$$

The only nonzero contribution to Q is given by the term with $q = 2$ and $r = 0$. Since $0 \leq r \leq M_{\sigma'} - 1$, the sufficient condition for $W_\alpha^{M_\sigma} = 0$ is $3 \leq q \leq M_\sigma - M_{\sigma'} + 1$. Therefore, we have for $M_\sigma \geq M_{\sigma'}$

$$Q = \begin{cases} M_{\sigma'} & \text{for } q = 2, \\ 0 & \text{for } 3 \leq q \leq M_\sigma - M_{\sigma'} + 1. \end{cases} \quad (4.B.93)$$

Q.E.D.

Chapter 5

Strings in Long-Range Interaction Model

5.1 Introduction

The $SU(2)$ Heisenberg spin chain with inverse-square exchange (ISE) [40] has given many remarkably simple non-trivial exact results. The exact ground state wave function of the model, for example, is the Jastrow-product or 1D analog of the Kalmeyer-Laughlin wave function [60]. The exact spin-spin correlation function [40] and thermodynamics [45] are also known. I show in this chapter another remarkable exact result. The "string hypothesis" originally proposed for the Bethe-ansatz solvable, nearest neighbor exchange (NNE) model in the thermodynamic limit (i.e. $N \to \infty$) is exact for the *finite-size* ISE model. The complete $SU(n)$ highest weight eigenstates for the $SU(n)$ ISE model [31, 67] using the string occupation-number configuration and, thereby, provide the full representation content of the Yangian highest weight states [42] of the model. The exact thermodynamics is also obtained.

It is well known that the spin rapidities of the Bethe ansatz equations for the S=1/2 Heisenberg chain with the NNE form "strings" in the thermodynamic limit as shown in previous chapters. (Reminder: A set of complex rapidities that share a common real part and are separated along the imaginary axis by $\sqrt{-1}$ is called a string.) The exact numerical solution of the Bethe ansatz equation [41] shows that the strings are somewhat deformed for the finite-size NNE chain. However, there is one-to-one correspondence between the states given by the strings and the true $S = S_z$ eigenstates (i.e. the SU(2) highest weight states) of the chain. It has already been suggested by Inozemtsev [55] that the NNE model and the ISE model are two different singular limits of a more general integrable model. From a more detailed study of the Inozemtsev model [41] one can find that for the ISE model the strings are completely "squeezed" onto the real axis so that there is no "room" for any distortion of the strings due to the finite size effect.

5.2 Yangian Algebra

The underlying quantum symmetry algebra of the ISE model is called the Yangian algebra [42, 16, 8]. And, the correct representation content of the Yangian highest-weight states (YHWS) of the ISE model is given by the string description [31].

The Yangian (algebra) has been introduced by Drinfeld [16] and has further been explored by Bernard and Felder [8]. The Yangian $Y(\mathcal{G})$ over the semi-simple Lie algebra $\mathcal{G}$ is a special case of a Hopf algebra and is developed as an algebraic method for constructing solutions of the quantum Yang-Baxter equation.

I follow the notations given in [42] for the introduction of the Yangian. First, let me give the basics of Lie algebra. For more details please refer to Ref. [15]. If the traceless Hermitian matrices $\mathbf{t}^a$ are the fundamental representation of the generators of $SU(n)$ Lie group, then $[\mathbf{t}^a, \mathbf{t}^b] = f^{abc}\mathbf{t}^c$ and let $\text{Tr}(\mathbf{t}^a\mathbf{t}^b) = \frac{1}{2}\delta^{ab}$. The antisymmetric structure constant satisfies $f^{abc} = c_v f_{abc}$ and $f^{abc}f_{bcd} = \delta^a_d$ with $c_v = n$. An element $X_i^{\alpha\beta}$ of the $n \times n$ operator matrix $\mathbf{X}_i$ acts as $|\beta\rangle\langle\alpha|$ on site i, and the local $SU(n)$ spins are defined as $J_i^a = \text{Tr}(\mathbf{t}^a\mathbf{X}_i)$.

The global $SU(n)$ generators $Q_0^a = \sum_i J_i^a$ satisfy the usual Lie algebra

$$[Q_0^a, Q_0^b] = f^{abc}Q_0^c. \tag{5.1}$$

We call Q_0^a the level-0 generators of the Yangian. The level-n generators are obtained by the following recursive relation

$$f^{abc}[Q_1^c, Q_{n-1}^b] = c_v Q_n^a. \tag{5.2}$$

There are extra consistency conditions which the level-n generators must satisfy, and the lowest level non-trivial consistency relation is given by

$$[Q_0^a, [Q_1^b, Q_1^c]] - [Q_1^a, [Q_1^b, Q_0^c]] = A_{def}^{abc}\{Q_0^d, Q_0^e, Q_0^f\}, \tag{5.3}$$

where $A_{def}^{abc} = \frac{1}{4}h^2 f^{adk}f^{bel}f^{cfm}f_{klm}$ and $\{x_1, x_2, x_3\} = \sum_{i\neq j\neq k} x_i x_j x_k$. h is a normalization constant. If Eq. (5.3) is satisfied, all the higher level consistency conditions are satisfied automatically.

The Yangian is endowed with the homomorphisms $Q_n^a \to \Delta_+(Q_n^a)$ and $Q_n^a \to \Delta_-(Q_n^a)$ where the comultiplications $\Delta_\pm$ are given by

$$\Delta_\pm(Q_0^a) = Q_0^a \otimes 1 + 1 \otimes Q_0^a, \tag{5.4}$$

$$\Delta_\pm(Q_1^a) = Q_0^a \otimes 1 + 1 \otimes Q_0^a \pm \frac{1}{2}h f^{abc}Q_0^b \otimes Q_0^c. \tag{5.5}$$

The comultiplications act on two copies of the Hilbert space and encode how the generators act on the states in the Hilbert space. For example, in quantum mechanics the level-0 generator Q_0^a acts on the two-particle Hilbert space $H \otimes H$ as

$$Q_0^a(\psi_1 \otimes \psi_2) = Q_0^a\psi_1 \otimes \psi_2 + \psi_1 \otimes Q_0^a\psi_2, \tag{5.6}$$

where ψ denotes a state in the one-particle Hilbert space H. The action of the generators on the N particle Hilbert space is defined by the recursive applications of the action on the two particle space.

In the $SU(n)$ Cartan basis the generators are represented by $n - 1$ independent traceless Hermitian, diagonal matrices $\mathbf{h}^\mu$ ($\mu = 1, \ldots, n - 1$), and by $n(n - 1)$ non-Hermitian translation or shift operators $\mathbf{e}^{\mu,\nu}$ where $\mu \neq \nu$. If the level-n generators are represented in Cartan basis $\{H_n^\mu, E_n^{\mu,\nu}\}$, $E_n^{\mu,\mu+1}$ annihilates the Yangian highest weight states, i.e. $E_n^{\mu,\mu+1}|\Psi\rangle = 0$.

The level-1 generators of the Yangian for the ISE model are given by

$$Q_1^a(\{z_i\}) = {\sum_{ij}}' w_{ij} f^{abc} J_i^b J_j^c, \tag{5.7}$$

where $w_{ij} = (z_i + z_j)/(z_i - z_j)$. This generator satisfies the consistency condition Eq. (5.3). Therefore, the symmetry of the ISE model is fully described by the Yangian. The explicit form of the transfer matrix and the systematic way of generating the family of mutually commuting Hamiltonians as in the Bethe ansatz are found in [6, 98].

5.3   Squeezed Strings in the ISE model

We begin with the model Hamiltonian of the SU(n) "spin" system given by

$$H = 2 \sum_{i<j} J_{ij}(P_{ij} - 1) \tag{5.8}$$

where P_{ij} is an operator that exchanges the spins or states at sites i and j, and $J_{ij} = d(i - j)^{-2}$. $d(i - j)$ is the distance between sites and in the case of a ring of N sites $d(i - j) = (N/\pi)|\sin[\pi(i - j)/N]|$. If there are only two states available at each site the model is equivalent to $S = 1/2$ Heisenberg chain [40]. The SU(n) version of the model is given in [31, 67].

Haldane has constructed the "fully spin polarized spinon gas" (FPSG) states using the following Bethe-ansatz-like equation [45]

$$N k_i = 2\pi I_i + \pi \sum_j \text{sgn}(k_i - k_j) \tag{5.9}$$

where k_i is a pseudo-momentum assigned to each down spin and I_i the corresponding quantum number restricted to $|I_i| \leq (1/2)(N - M - 1)$ for M down spins. The unoccupied quantum numbers in the allowed set of $\{I_i\}$ represent the spin half objects called spinons always appearing in even numbers. The FPSG states with N_{sp} spinons have the total spin $S = N_{sp}/2$, thus the name FPSG. The spinons do not interact in the ISE model and, therefore, the degeneracy of the state is simply $2S + 1$. This idea has been generalized to SU(n) case in [42] and is discussed later in this chapter. So far, we have been able to set an upper bound to the correct representation content of the SU(n) analog of the FPSG states. More work in this direction is needed.

A different formulation that gives the correct representation content of the SU(n) analog of the FPSG states is given explicitly using the strings. [32] Let us begin with the SU(2) case and propose the follow string equation

$$N k_{i,\nu} = 2\pi I_{i,\nu} + \pi \sum_{j,\mu} c_{\nu\mu} \text{sgn}(k_{i,\nu} - k_{j,\mu}) \tag{5.10}$$

$$c_{\nu\mu} = \begin{cases} 2\text{Min}(\nu,\mu) & \text{for } \nu \neq \mu, \\ 2\nu - 1 & \text{for } \nu = \mu \end{cases} \tag{5.11}$$

where $k_{i,\nu}$ is the "real part" of pseudo-momenta belonging to ith string with "length" ν.[1] $\mathrm{Min}(\nu, \mu) = \nu$ if $\nu \leq \mu$, and $= \mu$ otherwise. We also impose the following condition: $k_{i,\nu} \neq k_{j,\mu}$ for $(i, \nu) \neq (j, \mu)$. Even though all the strings are "squeezed" onto the real axis we distinguish the strings with different length. The total energy and crystal momentum are $E = (1/2) \sum_{i,\nu}(k_{i,\nu}^2 - \pi^2)$ and $P = \sum_{i,\nu}(k_{i,\nu} + \pi)$ (mod 2π). The crystal momentum has the form, $P = 2\pi Q/N$, where Q is an integer restricted to $0 \leq Q < N$. We can also set $-N/2 < Q \leq N/2$ (or $-\pi < P \leq +\pi$) by taking $Q - N$ for $Q > N/2$. We have chosen the energy for the ferromagnetic state to be zero. Note that if we restrict to only 1-strings Eq.(5.10) reduces to Eq.(5.9). The quantum number $I_{i,\nu}$ is restricted to $|I_{i,\nu}| \leq (1/2)(N - \sum_{\mu} c_{\nu\mu}\alpha_{\mu} - 1)$, where α_{μ} is the number of strings with length μ. Hence, the total number of string solutions for N total spins with M down spins is given by

$$n(N, M) = \sum_{\{\alpha_{\nu}\}} \prod_{\nu} \binom{N - \sum_{\mu=1}^{M} c_{\nu\mu}\alpha_{\nu}}{\alpha_{\nu}}, \tag{5.12}$$

where the sum extends over all possible configurations $\{\alpha_{\nu}\}$ such that $\sum_{\nu=1}^{M} \nu\alpha_{\nu} = M$. This sum can be carried out to yield $\binom{N}{M} - \binom{N}{M-1}$. In analogy with the Bethe ansatz the states given by the string solutions of the Eq. (5.10) correspond to the SU(2) highest weight states. Since the Hamiltonian is SU(2) invariant the essential degeneracy for the state with $S = N/2 - M$ would be $N - 2M - 1$. And the total number of states is given by $\sum_{i=0}^{[N/2]} n(N, i)(N - 2i - 1) = 2^N$ where $[x]$ is equal to the greatest integer $\leq x$. Hence, the string solution is complete.

We checked the validity of the string description using the exact numerical diagonalization and show the result for N=6 chain in Table 5.1. The total crystal momentum is given by $2\pi Q/N$ where Q is an integer such that $-N/2 < Q \leq N/2$. The allowed values of Nk/π for this chain is $(-6, -4, -2, 0, 2, 4, 6)$ with the first and the last momenta always unoccupied. An eigenstate is described by the string occupation-number configuration. For example, the 2 and 1 in the string configuration, 0020010, correspond to a 2-string with $k = -\pi/3$ and a 1-string with $k = 2\pi/3$, respectively. The zeros correspond to unoccupied momenta.

The state with $E = -16$ and the irreducible Young's diagram (IYD) given by $[1^2]$ in Table 5.1 is peculiar. The quantum numbers for the 2-string and 1-string are both zero and the two configurations 0020100 and 0010200 both give the same energy and momentum. The negative momentum states are given by the mirror reflections of string configurations of the corresponding positive momentum states as shown in Table 5.1 and 5.2. The state mentioned above has no corresponding negative momentum state (i.e. $Q = 0$ and *both* of the two quantum numbers are zero) and yet the string configuration is not mirror-reflection invariant. The mirror-reflection invariance is restored by representing the state with both of the two possible configurations.

---

[1] The "length" here means the total number of pseudo momenta included in the string parameterized by $k_{i,\nu}$. We shall call a string with length ν the ν-string.

# of states	$|Q|$	E	Degeneracy	IYD	String Configuration
1	3	-19	1	$[1^2]$	0101010
4	0	-16	3	$[2^1]$	0010100
5	0	-16	1	$[1^2]$	0010200 or 0020100[a]
11	2	-14	3	$[2^1]$	0001010(0101000)
17	1	-13	3	$[2^1]$	0010010(0100100)
19	1	-13	1	$[1^2]$	0020010(0100200)
22	0	-10	3	$[2^1]$	0100010
27	3	-9	5	$[4^1]$	0001000
30	3	-9	3	$[2^1]$	0002000
31	3	-9	1	$[1^2]$	0003000
41	2	-8	5	$[4^1]$	0000100(0010000)
47	2	-8	3	$[2^1]$	0000200(0020000)
57	1	-5	5	$[4^1]$	0000010(0100000)
64	0	0	7	$[6^1]$	0000000

Table 5.1. Energy Spectrum for N=6 SU(2) chain. The momentum is defined as $2\pi Q/N$. The energy E is in units of $(2\pi/N)^2$. The irreducible Young's diagrams (IYD) for all the states are also listed. In the string configuration 0 is for unoccupied momenta, 1 for 1-string, 2 for 2-string and so on in the possible momenta $\pi(-1, -2/3, -1/3, 0, 1/3, 2/3, 1)$. The configurations in parenthesis are the corresponding negative momentum states.

The FPSG states correspond to the states with only 1-strings. The non-FPSG states, in general, contain r-strings with "length" greater than one. All of the non-FPSG states are found to be degenerate to the FPSG states. The energy depends explicitly only on the location of the strings and not on the length of the strings (see Table 5.1 and 5.2 for examples). This explains why the states with only 1-strings (i.e. FPSG) exhaust all the possible eigenenergies of the system. The hidden symmetry responsible for the large number of degeneracy is the Yangian symmetry [42]. The states with only 1-strings (i.e. FPSG states) are the reference states corresponding to the YHWS.

It is straightforward to generalize the SU(2) case to SU(n) following Sutherland [96]. Let us first examine SU(3) case. We let $N = m_1 + m_2 + m_3$ where m_i is the number of ith spin or state and let $m_1 \geq m_2 \geq m_1$, $M_{(1)} = m_2 + m_3$, $and M_{(2)} = m_3$. In this case we have the following string equations

$$N k_{i,\nu}^{(1)} = 2\pi I_{i,\nu}^{(1)} - \pi \sum_{j,\mu} \text{Min}(\nu,\mu)\text{sgn}(k_{i,\nu}^{(1)} - k_{j,\mu}^{(2)}) + \pi \sum_{j,\mu} c_{\nu\mu}\text{sgn}(k_{i,\nu}^{(1)} - k_{j,\mu}^{(1)}) \quad (5.13)$$

$$\pi \sum_{j,\mu} \text{Min}(\nu,\mu)\text{sgn}(k_{i,\nu}^{(2)} - k_{j,\mu}^{(1)}) = 2\pi I_{i,\nu}^{(2)} + \pi \sum_{j,\mu} c_{\nu\mu}\text{sgn}(k_{i,\nu}^{(2)} - k_{j,\mu}^{(2)}) \quad (5.14)$$

We have introduced $M_{(1)}$ and $M_{(2)}$ pseudo-momenta $k^{(1)}$ and $k^{(2)}$, respectively. The total energy and momentum depend explicitly only on $k^{(1)}$ and are given by $E = \sum_{i,\nu}(1/2)((k_{i,\nu}^{(1)})^2 - \pi^2)$ and $P = \sum_{i,\nu}(k_{i,\nu}^{(1)} + \pi)$ (mod 2π). The two sets of quantum numbers $\{I_n^{(1)}\}$ and $\{I_n^{(2)}\}$ are restricted to

$$|I_\nu^{(1)}| \leq \frac{1}{2}\left(N + \sum_\mu \text{Min}(\nu,\mu)\alpha_{\mu,(2)} - \sum_\mu c_{\nu\mu}\alpha_{\mu,(1)} - 1\right) \quad (5.15)$$

$$|I_\nu^{(2)}| \leq \frac{1}{2}\left(\sum_\mu \text{Min}(\nu,\mu)\alpha_{\mu,(1)} - \sum_\mu c_{\nu\mu}\alpha_{\mu,(2)} - 1\right) \quad (5.16)$$

where $\alpha_{\nu,(i)}$ is the number of ν-strings of type i. Hence, the total number of string solutions is

$$n(N, M_{(1)}, M_{(2)}) = \sum_{\{\alpha_{\nu,(1)}\},\{\alpha_{\mu,(2)}\}} \prod_{\nu,\mu} \binom{N + \sum_{\gamma=1}^{M_{(2)}} \text{Min}(\nu,\gamma)\alpha_{\gamma,(2)} - \sum_{\gamma=1}^{M_{(1)}} c_{\nu\gamma}\alpha_{\gamma,(1)}}{\alpha_{\nu,(1)}}$$

$$\times \binom{\sum_{\gamma=1}^{M_{(1)}} \text{Min}(\nu,\gamma)\alpha_{\gamma,(1)} - \sum_{\gamma=1}^{M_{(2)}} c_{\mu\gamma}\alpha_{\gamma,(2)}}{\alpha_{\mu,(2)}}. \quad (5.17)$$

We have not been able to explicitly evaluate $n(N, M_{(1)}, M_{(2)})$. However, we checked numerically for $N \leq 15$ that

$$\sum_{m_1 \geq m_2 \geq m_3} n(N, M_{(1)}, M_{(2)})\nu(m_1, m_2, m_3) = 3^N \quad (5.18)$$

where $\nu(m_1, m_2, m_3)$ is the essential degeneracy of the SU(3) invariant state and is given by $(m_1 - m_2 + 1)(m_2 - m_3 + 1)(m_1 - m_3 + 2)/2$. Hence, we conclude that the string solution is complete. We also checked with the exact numerical diagonalization of SU(3) system of size up to 9. In Table 5.2 we enumerate the string states for N=6. For the lack of space we list only lowest 178 states.

For SU(n) case we propose the following set of string equations:

$$Nk_{i,\nu}^{(1)} = 2\pi I_{i,\nu}^{(1)} - \pi \sum_{j,\mu} \text{Min}(\nu,\mu)\text{sgn}(k_{i,\nu}^{(1)} - k_{j,\mu}^{(2)}) + \pi \sum_{j,m} c_{\nu\mu}\text{sgn}(k_{i,\nu}^{(1)} - k_{j,\mu}^{(1)}) \quad (5.19)$$

$$\pi \sum_{j,\mu} \text{Min}(\nu,\mu)\text{sgn}(k_{i,\nu}^{(r)} - k_{j,\mu}^{(r-1)}) + \pi \sum_{j,\mu} \text{Min}(\nu,\mu)\text{sgn}(k_{i,\nu}^{(r)} - k_{j,\mu}^{(r+1)}) =$$

$$2\pi I_{i,n}^{(r)} + \pi \sum_{j,m} c_{nm}\text{sgn}(k_{i,n}^{(r)} - k_{j,m}^{(r)}) \quad (5.20)$$

where $r = 2, 3, \ldots, n-1$. We define $\text{sgn}(k_{i,\nu}^{(n)} - k_{j,\mu}^{(n+1)}) \equiv 0$. If only 1-strings are allowed the above equations reduce to the equations obtained in [67]. The total energy and crystal momentum have the same form as in the SU(3) case. The quantum numbers are restricted to

$$|I_\nu^{(1)}| \leq \frac{1}{2}\left(N + \sum_\mu \text{Min}(\nu,\mu)\alpha_{\mu,(2)} - \sum_\mu c_{\nu\mu}\alpha_{\mu,(1)} - 1\right) \quad (5.21)$$

$$|I_\nu^{(r)}| \leq \frac{1}{2}\left(\sum_\mu \text{Min}(\nu,\mu)\alpha_{\mu,(r-1)} + \sum_\mu \text{Min}(\nu,\mu)\alpha_{\mu,(r+1)} - \sum_\mu c_{\nu\mu}\alpha_{\mu,(r)} - 1\right) \quad (5.22)$$

where $r = 2, 3, \ldots, n-1$ and $\alpha_{\mu,(n)} = 0$.

The full SU(n) representation content of a YHWS is simply given by all the possible string configurations whose occupied momenta are identical to that of the YHWS. Eq. (5.21) and Eq. (5.22) are the "string selection rules" that screen out the configurations that do not belong to the representation content of the YHWS.

5.4 Construction of Motifs

The idea of *motif* introduced in [42] is useful for exhibiting the coproduct structure of the YHWS. As discussed in [42] the complete set of eigenvalues for the $SU(n)$ chain are given by the string configurations containing only 1-strings such that *no more than n − 1 consecutive 1-strings appear*. The string configuration, for example, for the $N = 6$ $SU(3)$ singlet ground state is given by 0110110 as shown in Table II. We replace the first and the last 0's by the symbols "(" and ")", respectively. All other zeros acquire a full symbol ")(". Hence, the string configuration 0110110 is represented by the product of motifs "(11)(11)." "(11)" is the $SU(3)$ ground state motif. The $SU(n)$ ground state motif would be $(1\ldots1)$ with $n-1$ consecutive 1's. The excited states in general contain (), (1), (11), and so on up to the ground state motif. The symbols "(" and ")" each have a value of 1/2. The "1" has a full value of 1. For example, the motif (11) has a value of 3. A motif can be represented with a Young's diagram $[1^m]$ where m is the total symbol counts of the motif. (The Young's diagram

# of states	$|Q|$	E	Degeneracy	IYD	String Configuration(for $k^{(1)}$)
1	0	-26	1	$[1^3]$	0110110
17	2	-22	8	$[2^1 1^1]$	0011010(0101100)
19	2	-22	1	$[1^3]$	0021010(0101200)
35	1	-21	8	$[2^1 1^1]$	0010110(0110100)
45	3	-19	10	$[3^2]$	0101010
53	3	-19	8	$[2^1 1^1]$	0101010
61	3	-19	8	$[2^1 1^1]$	0101010
62	3	-19	1	$[1^3]$	0102010
78	2	-18	8	$[2^1 1^1]$	0100110(0110010)
98	1	-17	10	$[3^1]$	0001100(0011000)
114	1	-17	8	$[2^1 1^1]$	0002100(0012000)
141	0	-16	27	$[4^1 2^1]$	0010100
151	0	-16	10	$[3^1]$	0010100
161	0	-16	10	$[3^2]$	0020100 or 0010200[a]
169	0	-16	8	$[2^1 1^1]$	0020100
177	0	-16	8	$[2^1 1^1]$	0010200
178	0	-16	1	$[1^3]$	0020200

Table 5.2. Energy Spectrum for N=6 SU(3) Chain. [a] The quantum numbers for the 2- and 1-string are both zero and the two configurations are possible.

$[1^m]$ corresponds to a single column of m boxes.) The direct product of the irreducible Young's diagrams gives the upper bound to the correct $SU(n)$ representation content of the YHWS. The mirror-reflection of configuration representing a YHWS naturally has the same spin content.

By examining the configurations given by Eq.(5.19)-(5.22), we found the following *fusion* rule that reduces the upper bound given by the direct product of the Young's diagrams: *a string of Young's diagram $[1^1]$ representing the motif () and one additional Young's diagram representing a non-() motif on the far left side of the string of ()'s combine only in a* **totally symmetric** *way.* For example, the string of motifs $(1)()()()()$ is represented by $[5^1 1^1]$. The YHWS represented by 0001100 in Table II has the motif representation $()()(11)()$. So, the upper limit of the $SU(3)$ representation content of the state is given by $[2^1] \otimes [1^2] = [3^1] \oplus [2^1 1^1]$. In this case the upper limit happens to be the exact $SU(3)$ spin content.

Additional rules are buried in Eq.(5.19)-(5.22). We give one more rule as follow. From the string equations we can easily show that the first level pseudo-momentum of a ν-string satisfies the following condition

$$|k_\nu^{(1)}| \le (\pi/N)(N - 2\nu). \tag{5.23}$$

Hence, the momenta $\pm\pi$ are always unoccupied, and the longer strings are driven towards the center of a configuration (i.e. $k = 0$). We give here an example to illustrate how we can use this new rule to eliminate further the extra states obtained by the direct product of the Young's diagrams contained in the YHWS. The YHWS with $E = -13$ and $|Q| = 1$ is represented by the 1-string configuration 0100100 or by the motifs $(1)()(1)()$. By the fusion rule the two motifs () are "fused" onto each of the two neighboring motifs (1). Hence, the upper limit on the full $SU(3)$ spin content is given by the following direct product: $[2^1 1^1] \otimes [2^1 1^1] = [4^1 2^1] \oplus [3^2] \oplus [3^1] \oplus [2^1 1^1] \oplus [2^1 1^1] \oplus [1^3]$. (Fig. 5.1 shows this direct product pictorially.) The actual spin content of the YHWS is $[4^1 2^1] \oplus [3^2] \oplus [3^1] \oplus [2^1 1^1]$, and the states represented by one of the two $[2^1 1^1]$'s and $[1^3]$ are apparently forbidden. We can explain this exclusion easily from the rule given by Eq. (5.23) as follow. The string configurations corresponding to the IYD $[2^1 1^1]$ are 0100200 (0020010) and 0200100 (0010020). The configuration 0200100 violates the rule given by Eq. (5.23) since the 2-string can occupy only the 3rd, 4th or the 5th position according to the rule. For this reason the state represented by one of $[2^1 1^1]$'s does not appear in the multiplet. For the same reason, the configuration 0200200 representing $[1^3]$ is forbidden and does not appear in the multiplet.

5.5   Thermodynamics

The exact thermodynamics can also be obtained by the standard method [110, 25, 99]. Before taking the limit $N \to \infty$ for the string equations (5.19)-(5.22), we subtract the ith from the $(i + 1)$th equations and divide the both sides by $N(k_{i+1,\nu}^{(r)} - k_{i,\nu}^{(r)})$. The resulting equations should contain terms like $\frac{1}{N}(I_{i+1,\nu}^{(r)} - I_{i,\nu}^{(r)})/(k_{i+1,\nu}^{(r)} - k_{i,\nu}^{(r)})$ which

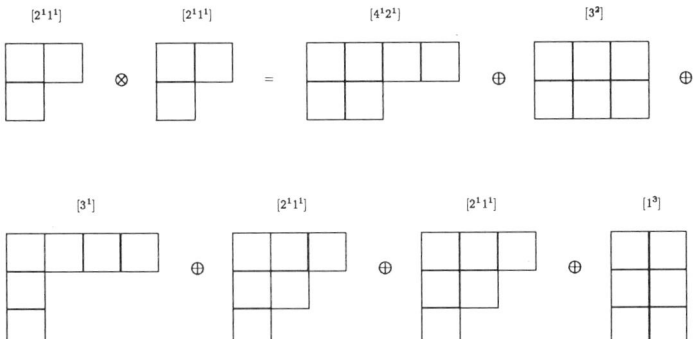

Fig. 5.1. Direct product $[2^1 1^1] \otimes [2^1 1^1]$ is expressed as a direct sum of the $SU(3)$ Young's diagrams which are labeled by the corresponding irreducible diagrams.

can be thought of as a density function for the k as $N \to \infty$. This density function is indeed a sum of the density functions (ρ and ρ^h) for the k's that correspond to the occupied quantum numbers ("particles") and for the unoccupied ones ("holes"). After some algebraic manipulations we obtain the following set of coupled equations for the particle and hole density functions

$$\frac{1}{2\pi} = \rho_\nu^{(1)}(k) + \rho^{h(1)}_\nu(k) - \sum_\mu \min(\nu, \mu) \rho_\mu^{(2)}(k) + \sum_\mu c_{\nu\mu} \rho_\mu^{(1)}(k) \tag{5.24}$$

$$\rho_\nu^{(r)} + \rho^{h(r)}_\nu = \sum_\mu \min(\nu, \mu) \rho_\mu^{(r-1)}(k) - \sum_\mu c_{\mu\nu} \rho_\mu^{(r)}(k) + \sum_\mu \min(\nu, \mu) \rho_\mu^{(r+1)}(k) \tag{5.25}$$

where $r = 2, 3, \ldots, n-1$ and $\rho^{(n)} \equiv 0$. There are infinite number of solutions that satisfy the above equations. We would like to find an equilibrium density functions at some fixed temperature T and express the free energy in terms of those density functions.

The free energy is $F = E - TS - \sum_r h'_r n_r$, where h'_r is the magnetic field projected onto the r-spin species and n_r the number of r spins. S is the entropy and can be obtained as follow. The total number of particles and holes in the infinitesimal region $(k, k + dk)$ is $N\rho(k)dk$ and $N\rho^h(k)dk$, respectively. So, the entropy is given by

$$\ln \left\{ \frac{[(\rho + \rho^h) N dk]!}{[N\rho dk]! [N\rho^h dk]!} \right\} \approx \left[(\rho + \rho^h) \ln(\rho + \rho^h) - \rho \ln \rho - \rho^h \ln \rho^h \right] N dk \tag{5.26}$$

We, of course, need to sum over the spin and the string index to obtain the total entropy. Finally, in terms of the density functions the free energy is

$$F/N \equiv f = \sum_\nu \int dk \rho_\nu^{(1)}(k) e(k) - T \sum_{i,\nu} \int dk \left\{ (\rho_\nu^{(i)} + (\rho_\nu^{(i)})^h) \ln(\rho_\nu^{(i)} + (\rho_\nu^{(i)})^h) \right.$$

$$- \rho_\nu^{(i)} \ln \rho_\nu^{(i)} - (\rho_\nu^{(i)})^h) \ln(\rho_\nu^{(i)})^h) \Big\} - \sum_{i,\nu} \int dk h_i \rho_\nu^{(i)}, \qquad (5.27)$$

where $e(k) = (1/2)(k^2 - \pi^2)$. One can check that h_i is trivially related to h_i'.

We can obtain the equilibrium distribution function by setting $\delta f[\{\rho\}, \{\rho^h\}] = 0$ and solving the resulting expression subjected to the conditions (5.24)- (5.25). In terms of the newly introduced density function $\eta = \rho^h/\rho$, the free energy per site is given by

$$f = -T \sum_{n=1}^\infty \int_{-\pi}^\pi \frac{dk}{2\pi} \ln(1 + (\eta_n^{(1)})^{-1}), \qquad (5.28)$$

where η_n is determined by the following set of equations:

$$(\eta_n^{(\nu)})^2 = \frac{(1 + \eta_{n-1}^{(\nu)})(1 + \eta_{n+1}^{(\nu)})}{(1 + (\eta_n^{(\nu-1)})^{-1})(1 + (\eta_n^{(\nu+1)})^{-1})}, \qquad (5.29)$$

where $\nu = 1, 2, 3, \ldots, s - 1$ and $\eta^{(0)}, \eta^{(s)} = 0$. The boundary conditions are

$$\eta_0^{(r)} = (\exp[e(k)/T] - 1)\delta_{r,1}, \qquad (5.30)$$

$$\lim_{n \to \infty} \ln(\eta_n^{(r)})/n = 2h_r/T. \qquad (5.31)$$

The thermodynamic equations for the $SU(2)$ case has also been obtained in [43] by taking the appropriate singular limit of the corresponding equations for the NNE model. Our work shows that the limit used in [43] is equivalent to squeezing the strings to the real axis.

5.6 Conclusion

I showed in this chapter that the strings originally introduced for the nearest-neighbor Bethe-ansatz solvable models give the exact and complete $SU(n)$ highest weight states of the *finite-size* ISE model and found that the strings give the complete multiplet structure of the YHWS. [32] It is interesting to note that the phase shifts $c_{\mu\nu}$ code the exclusion statistics for the strings.

The $SU(n)$ Heisenberg model can be considered as $SU(n-1)$ supersymmetric bosonic t-J model by regarding a spin species as holes. It is obvious that the $SU(n)$ supersymmetric fermionic t-J model discussed in Chapter 3 can also be solved using the same method presented in this paper. The NNE Hubbard model discussed in Chapter 2 may have a corresponding long-range interaction model whose strings are squeezed.

Chapter 6

Elementary Excitations of t-J Model

6.1 Introduction

In this chapter I discuss exact excitation contents of the intermediate states for the one-particle Green's functions, spin-spin and (charge) density-density correlation functions of the periodic one-dimensional t-J model with inverse square exchange [33]. The excitations consist of neutral $S = 1/2$ spinons and spinless (charge $-e$) holons with semionic fractional statistics, and bosonic (charge $+2e$) "anti-holons" which are excitations of the holon condensate. A set of selection rules and the regions of non-vanishing spectral weight in the energy-momentum space for the various correlation functions are found.

As discussed in previous two chapters many interesting and simple features in the family of Calogero-Sutherland models (CSM) are identified with their peculiar inverse-square exchange (ISE) [11, 95, 40, 42, 70, 31, 67, 55]. Most important feature of these models is that they belong to the same low-energy universality class as the family of Bethe-ansatz solvable models and perhaps even larger family of integrable systems and may provide a new *fully* soluble paradigm next to the non-interacting models [45].

The one-dimensional supersymmetric ISE t-J model [70] represents a fixed point model where the elementary excitations form an ideal gas obeying fractional statistics. In contrast to this model, the NNE t-J model [96, 3], which has essentially the same low energy spectra spanned by the same elementary excitations, obscures the simple low energy structure intrinsic to this class of models. The spinons, the holons and the antiholons (the elementary excitations of the NNE t-J model [3]) are rediscovered in the context of the supersymmetric Yangian of the ISE model. Furthermore, it is found that only the states with *finite* number of these elementary excitations contribute to the spectral functions of the one-particle Green's functions ($G^{(1)}$), the charge density-density ($C^{(c)}$) and the spin-spin correlation functions ($C^{(s)}$).

6.2 Supersymmetric t-J Model

First, we examine the symmetry in the ISE supersymmetric t-J model. The model with periodic boundary conditions possesses, in addition to the global $SU(m|n)$ supersymmetry, a hidden dynamical "quantum group" symmetry algebra called the supersymmetric Yangian [45, 42, 16]. This symmetry is responsible for the "supermultiplets" in the energy spectrum and the ideal gas-like features of the elementary excitations and, furthermore, provides us with a simple numerical way to identify the exact content of the elementary excitations relevant for the various correlation functions.

The supersymmetric generalization of the $SU(n)$ Haldane-Shastry model Hamiltonian [46, 31, 67] is given by

$$H = t \sum_{i<j} \frac{P_{ij}}{d^2(n_i - n_j)}, \qquad (6.1)$$

where $d(x) = (N_a/\pi)\sin(\pi x/N_a)$ and N_a is the total number of sites. If $a_{i\alpha}^{\dagger}$ ($a_{i\alpha}$) creates (destroys) a particle of species α at site i and satisfies the single occupancy condition, $\sum_{\alpha} a_{i\alpha}^{\dagger} a_{i\alpha} = 1$, the exchange operator can be written as $P_{ij} = \sum_{\alpha\beta} a_{i\alpha}^{\dagger} a_{j\beta}^{\dagger} a_{i\beta} a_{j\alpha}$. If m of the species labeled by α are bosons, and n are fermions, the model (6.1) has a global $SU(m|n)$ supersymmetry with generators given by the traceless part of $J_0^{\alpha\beta} = \sum_i a_{i\alpha}^{\dagger} a_{i\beta}$. The Yangian symmetry generator of the periodic ISE model is

$$J_1^{\alpha\beta} = \sum_{i>j,\gamma} w_{ij} a_{i\alpha}^{\dagger} a_{j\gamma}^{\dagger} a_{i\gamma} a_{j\beta}, \qquad (6.2)$$

where $w_{ij} = \cot(\pi(i-j)/N_a)$. The higher order generators of the Yangian are obtained recursively from various commutators involving only J_0 and J_1 [42, 16].

If we specialize to $SU(1|2)$ supersymmetry, with $\alpha \in \{0, \uparrow, \downarrow\}$, we can rewrite the Hamiltonian in terms of the $SU(2)$ fermionic operators $c_{i\sigma}^{\dagger} = a_{i\sigma}^{\dagger} a_{i0}$ as $\mathcal{P}H^0\mathcal{P}$, where H^0 (up to a shift in total energy and in chemical potential) is

$$- \sum_{i \neq j,\sigma} t_{ij} c_{i\sigma}^{\dagger} c_{j\sigma} + \sum_{i<j} (J_{ij}\mathbf{S}_i \cdot \mathbf{S}_j + V_{ij}n_in_j), \qquad (6.3)$$

where $t_{ij} = J_{ij}/2 = -2V_{ij} = t/d^2(i-j)$ and $n_i = n_{i\uparrow} + n_{i\downarrow}$; $\mathcal{P}$ is the projection operator that projects out all states with doubly-occupied sites. The ground state $|\Psi_0\rangle$ of this model is known [70, 31] to be

$$\sum_{\{x,\sigma\}} \prod_{i>j} (z_i - z_j)^{\delta_{\sigma_i,\sigma_j}} (i)^{\text{sgn}(\sigma_i-\sigma_j)} \prod_k z_k^{J_0} \prod_j c_{x_j\sigma_i}^{\dagger} |0\rangle, \qquad (6.4)$$

where $z_j = \exp(i2\pi x_j/N_a)$, $J_0 = -(N/2 - 1)/2$, N is the total number of particles, and $|0\rangle$ the electron vacuum (empty state). In order to have a non-degenerate ground state one might take $N/2$ to be odd. Note that this wavefunction has an implicit projection that makes the wavefunction vanish whenever the configuration contains multiply-occupied sites and, therefore, is just the fully Gutzwiller projected free electron state[85]. The wavefunction can also be written in terms of the M_h hole and $M_\downarrow$ down-spin particle locations as follow

$$\prod_{\alpha>\beta}(Z_\alpha - Z_\beta)\prod_{\alpha,j}(Z_\alpha - Z_j^\downarrow)\prod_{i>j}(Z_i^\downarrow - Z_j^\downarrow)^2\prod_\alpha Z_\alpha^{J_h}\prod_j Z_j^{\downarrow J_\downarrow}, \qquad (6.5)$$

where Z's are the hole coordinates and $Z^\downarrow$'s the down-spin coordinates. Without any loss of generality let $M_\downarrow \le M_\uparrow$. Then, the normalization of the wavefunction is given by

$$|\Psi|^2 = N_a^{M_h+M_\downarrow}\prod_{j=1}^{M_\downarrow}(M_h + 2j - 1). \qquad (6.6)$$

Derivation of this normalization constant is given in Appendix 6.A

The form of wavefunction given in Eq. (6.5) suggests that the ground state is an analog of the hierarchical states in fractional quantum hall system [50, 91]. The last Jastrow term describes a bosonic Laughlin condensate and the second its hole excitations. The first Jastrow term describes the hierarchical condensate of the hole excitations. As will be shown later in this chapter this analogy is a quite accurate description of the 1D t-J model.

A remarkable feature of this model is that the eigenstates of (6.1) form degenerate "supermultiplets"[40, 45] with multiplicities much higher than those expected from the global supersymmetry. All supermultiplets on the $SU(m|n)$ model with $m, n > 0$ are present (with the same energy and momentum, but multiplicity reduced to 2) in the spinless free fermion $SU(1|1)$ model [46]. This means that they can be represented by a binary sequence of $N_a - 1$ ones and zeroes, representing (in the spinless fermion model) the occupations of Bloch states with non-zero momentum (the zero-momentum orbital has zero energy, which is the supersymmetry, and its occupation is not fixed). There are thus 2^{N_a-1} distinct supermultiplets.

6.3 Elementary Excitations

In the $SU(1|2)$ case, the "occupation number" sequence describes a supermultiplet spanning a large range of possible fermion charges N. The state of minimum charge in the supermultiplet is given by the number of zeroes in the sequence; the maximum charge is N_a minus the number of times two consecutive ones occur. The ground state of the model with $t > 0$ has a sequence $111\ldots 111$, so its minimum charge is $N = 0$ and its maximum charge is $N_a - (N_a - 2) = 2$. The multiplet represented by the alternating sequence $10101\ldots 10101$ has a maximum charge state $N = N_a$, which is the spin-singlet ground state of the antiferromagnetic $S = 1/2$ Haldane-Shastry chain, and a minimum charge $(N_a - 2)/2$.

We study the model (6.1) with $t > 0$ and a chemical potential that *maximizes* N, so the ground-state has $0 < N < N_a$. Then, only intermediate states with the *maximum* value of charge in their supermultiplet contribute to the thermodynamic limit of the ground-state correlation functions. To determine the excitation content of these maximal charge states, it is convenient to add a zero to each end of the binary sequence, expanding its length to $N_a + 1$. The ground state sequence is then of the form $0101010\ldots1111111\ldots0101010$, with a central section of consecutive ones, with equal-length wings of the alternating sequence.

In the limit $N = N_a$, the excitations of the $S = 1/2$ antiferromagnet are neutral spin-1/2 *spinons*[9, 20, 102] represented by consecutive zeroes (e.g. $\ldots0101001010\ldots$) and spinless charge $-e$ *holons* by consecutive ones (e.g. $\ldots010101101010\ldots$). At intermediate densities, the central region $\ldots1111111\ldots$ may be considered as a *holon condensate* or "pseudo-Fermi-sea". However, the holons and spinons are *not* fermions, but *semions*, or particles with "half-fractional" statistics, resulting from the spin-charge separation of a hole, which is a spin-1/2, charge $-e$ fermion. A configuration $\ldots11111110111111\ldots$ has a *"hole in the holon condensate"* which we will call an "antiholon"; because of the semionic statistics of the holons, we identify it as a charge $+2e$, spinless boson.

Using concepts from Chern-Simons theory, as applied to the fractional quantum Hall effect[112], if condensed particles have charge q and statistics $\Theta = \pi\lambda$, vortices or holes in the condensate have charge $-q/\lambda$ and statistics $\Theta' = \pi/\lambda$. Here holons have charge $-e$ and $\Theta = \pi/2$ (a semion), so the vortex or hole in the holon condensate (antiholon) then has charge $2e$ and $\Theta = 2\pi$ (a boson). The applicability of such "2D" concepts to 1D ISE-type models has recently been demonstrated: the holon (antiholon) corresponds to particle (hole) excitations of the $\lambda = 1/2$ Calogero-Sutherland model where the particle excitations are semions and the holes $\lambda = 2$ bosons [46, 34, 35].

The main results of this paper can be summarized in Table 6.1, which lists all the possible elementary excitations for the corresponding local perturbations of the ground state. The quantum symmetry prevents the injected electron or hole from breaking up into more than a very simple set of elementary excitations consisting of the left (right) spinons($s_{L(R)}$), holons($h_{L(R)}$), and antiholons($\bar{h}$). As a result, the spectral functions of the various dynamical correlation functions vanish except in certain regions of the energy-momentum plane (*i.e.*, has *"compact support"*).

6.4 Compact Supports of Various Correlation Functions

Figs. 6.1-6.3 show the regions of compact support formed by the *finite* number of elementary excitations contributing to the intermediate states for $G^{(1)}$, $C^{(c)}$, and $C^{(s)}$, respectively. If the correlation functions are given by the following integral,

$$C(x,t) = \int_{(Q,E)\in\sigma} dQ\, dE\, S(Q,E)\, e^{i(Qx-Et)}, \tag{6.7}$$

the figures show the region σ where the spectral function $S(Q,E)$ is non-zero; this is determined by combining the energies and (Bloch) momenta of the finite set of

Local Operator $\hat{O}_i$	Excitation contents of $\hat{O}_i|\Psi_0\rangle$
$c_{i\sigma}$	$(s_L, h_L) + \bar{h} + 2(s_R, h_R)$
$c_{i\sigma}^\dagger$	$(s_L, h_L) + \bar{h}$
$n_{i\uparrow} + n_{i\downarrow}$	$(s_L, h_L) + \bar{h} + (s_R, h_R)$
	$\bar{h} + 2h_R$
$n_{i\uparrow} - n_{i\downarrow}$	$(s_L, h_L) + \bar{h} + (s_R, h_R)$
	$2s_L$

Table 6.1. List of all possible excitations from the ground state perturbed by local operators $c_{i\sigma}(c_{i\sigma}^\dagger)$ $(G^{(1)})$, $n_{i\uparrow} + n_{i\downarrow}$ $(C^{(c)})$, and $n_{i\uparrow} - n_{i\downarrow}$ $(C^{(s)})$. The mirror states $(L \leftrightarrow R)$ not listed are also allowed. The spinon (v_s), holon (v_h), antiholon $(v_{\bar{h}})$, spin-wave (v_s^0) and sound (v_c^0) velocities always satisfy: (i) $v_c^0 < v_s^0$, (ii) $v_c^0 \leq |v_h|(|v_s|) \leq v_s^0$, (iii) $|v_{\bar{h}}| \leq v_c^0$, and (iv) for a given spinon-holon pair (s_R, h_R), $|v_{s_R}| \geq |v_{h_R}|$.

elementary excitations contributing to $S(Q, E)$.

The dispersion relations for the spinon, holon and antiholon in the thermodynamic limit are given respectively by

$$E_{s_{R(L)}}/t = -q(q \mp v_s^0), \tag{6.8}$$

$$E_{h_{R(L)}}/t = q(q \pm v_c^0), \tag{6.9}$$

$$E_{\bar{h}}/t = \frac{(v_c^0)^2 - q}{2}, \tag{6.10}$$

where $v_s^0 = \pi$ (spin-wave velocity), $v_c^0 = \pi(1 - \bar{n})$ (sound velocity) and $\bar{n}$ the density of electrons. The right (left) spinons and holons take the upper (lower) signs and are allowed only in $0 \leq q \leq \pi\bar{n}/2$ $(-\pi\bar{n}/2 \leq q \leq 0)$ relative to the $Q = 0$ ground state, while the antiholons propagate in region $-v_c^0 \leq q \leq v_c^0$. The curvature of the antiholon dispersion is half that of the holon, indicating that $\bar{h}$ is made by destroying two holons. It is then natural to assign charge $C = +2e$ and $S = 0$ to the antiholon while $C = 0$ and $S = \frac{1}{2}$ to the spinon, and $C = -e$ and $S = 0$ to the holon. This assignment is consistent with the results given in Table I and the phase shift calculations. In fact, using this charge conservation argument we were able to identify one extra right holon for the local hole excitation $(\hat{O}_i = c_{i\sigma})$ in Table I, which could not be resolved numerically because of the small system size $(N_a = 12)$ studied.

We outline below how to find the regions of support for the various correlation functions. First, we numerically find all the eigenstates having non-zero overlap with the states $c_{i\sigma}(c_{i\sigma}^\dagger)|\Psi_0\rangle$ (for $G^{(1)}$), $(n_{i\uparrow} + n_{i\downarrow})|\Psi_0\rangle$ (for $C^{(c)}$) and $(n_{i\uparrow} - n_{i\downarrow})|\Psi_0\rangle$ (for $C^{(s)}$). Second, we identify the excitation content of the states by inspecting the

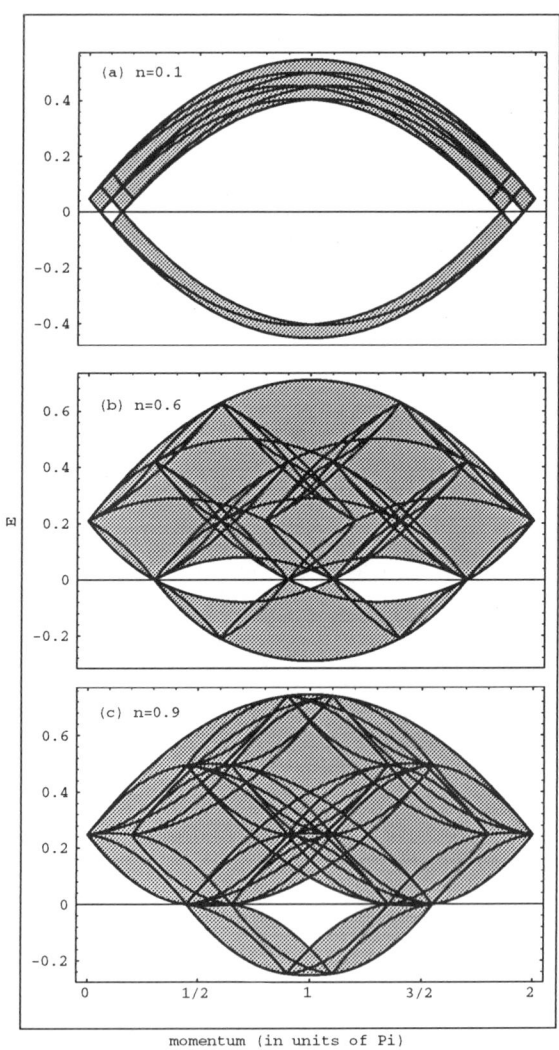

Fig. 6.1. *Compact support* for the one-particle Green's function. The momentum is in units of π and the excitation energy E in π^2/t. The contributing elementary excitations to this region are $(h_L, s_L) + \bar{h} + 2(h_R, s_R)$ for the positive energy part (i.e. $c_{i\sigma}|\Psi_0\rangle$) and $(s_L, h_L) + \bar{h}$ for the negative part (i.e. $c_{i\sigma}^\dagger|\Psi_0\rangle$). Their mirror states (i.e. L and R exchanged) also contribute. The four momenta at which $E = 0$ is allowed are k_F, $2\pi - 3k_F$, $3k_F$, and $2\pi - k_F$ where $k_F = \pi\bar{n}/2$.

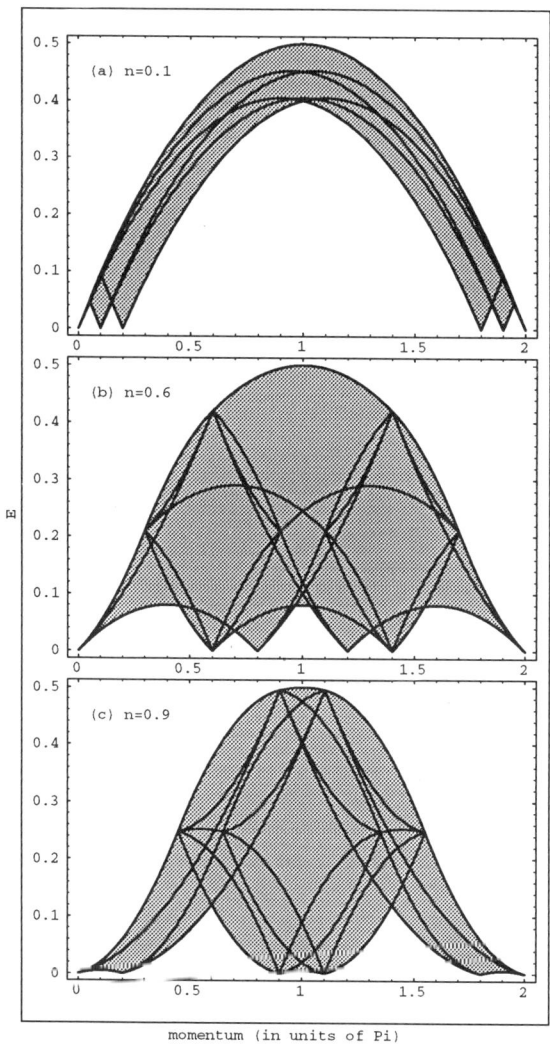

Fig. 6.2. *Compact support* for the density-density correlation function. $\left(s_L, h_L\right) + \bar{h} + \left(s_R, h_R\right)$, $\bar{h} + 2h_R$ and their mirror states contribute. $E = 0$ is allowed at $0(2\pi)$, $2k_F$, $2\pi - 4k_F$, $4k_F$, $2\pi - 2k_F$. Only holon-antiholon branches are present at $4k_F$ $(2\pi - 4k_F)$ indicating that $4k_F$ is the holon Fermi point.

corresponding motifs where the spinons, holons and antiholons can easily be identified (see Table I). We empirically find the following selection rules that the holon (v_h), spinon (v_s), antiholon $(v_{\bar{h}})$, spin wave (v_s^0) and sound (v_c^0) velocities always satisfy:

- (i) $v_c^0 < v_s^0$ (i.e. spin-charge separation),

- (ii) $v_c^0 \leq |v_h|(|v_s|) \leq v_s^0$,

- (iii) $|v_{\bar{h}}| \leq v_c^0$,

- (iv) for a given spinon-holon pair (s_R, h_R), $|v_{s_R}| \geq |v_{h_R}|$.

These rules together with the results in Table I allow us to plot the regions of compact support as shown in Figs. 6.1-6.3.

Fig. 6.1 shows the region of support for the one-particle Green's function where the states $c_{i\sigma}|\Psi_0\rangle$ $(c_{i\sigma}^\dagger|\Psi_0\rangle)$ propagate in time with positive (negative) energy with respect to the ground state. The spectral functions should be non-analytic along all the solid lines where the elementary excitations either "touch" the boundaries or the other elementary excitations. When the antiholons are suppressed (i.e. near half filling), the holon is accompanied either by a spinon or by three spinons in $S = 1/2$ state. At $3k_F$ $(2\pi - 3k_F)$, where $k_F = \pi\bar{n}/2$, the left (right) moving spinon is missing from the state $c_{i\sigma}^\dagger|\Psi_0\rangle$ since the charge conservation prevents more than one holon in the presence of one antiholon. Of course, if two antiholons were allowed then states of the type $(s_L, h_L) + 2\bar{h} + 2(s_R, h_R)$ would contribute. Our numerical study indicates that states with two antiholons do not contribute. In fact, the observed states listed in Table I are the simplest possible states satisfying the charge (spin) conservation with at most one antiholon.

In Fig. 6.2, only holon-antiholon branches are present at $4k_F$ $(2\pi - 4k_F)$ while the spinon-holon branches show up at $2k_F$ $(2\pi - 2k_F)$. In Fig. 6.3 we find that the pure spinon excitations are possible only if they both belong to the same sector, otherwise they are accompanied by two holons and an antiholon. The excitation content we find for $S_i^z(= (n_{i\uparrow} - n_{i\downarrow})/2)$ should be same for $S_i^\pm$ since the ground state is a spin singlet. As $\bar{n} \to 0$ we recover the two spinon spectrum for the $S = 1/2$ spin chain.

6.5 Interpolation of Anti-Holon Charge from NNE to ISE Model

Finally, we have examined how the ISE results for the charge of the elementary excitations change if we interpolate between the ISE and NNE $t - J$ models, which are respectively the $\gamma = 0$ and $\gamma = \infty$ limits of the integrable family of *hyperbolic* models with exchange $\propto 1/\sinh^2 \gamma(i - j)$[36]. Away from the ISE limit, the charge carried by the holon and antiholon excitations vary with their velocity, and become equal in magnitude (and opposite in sign) as the velocities approach the sound velocity v_c^0. In the ISE limit, however the holon charge $(|v| > v_c^0)$ is always $-e$, and the antiholon charge $(|v| < v_c^0)$ is always $+2e$.

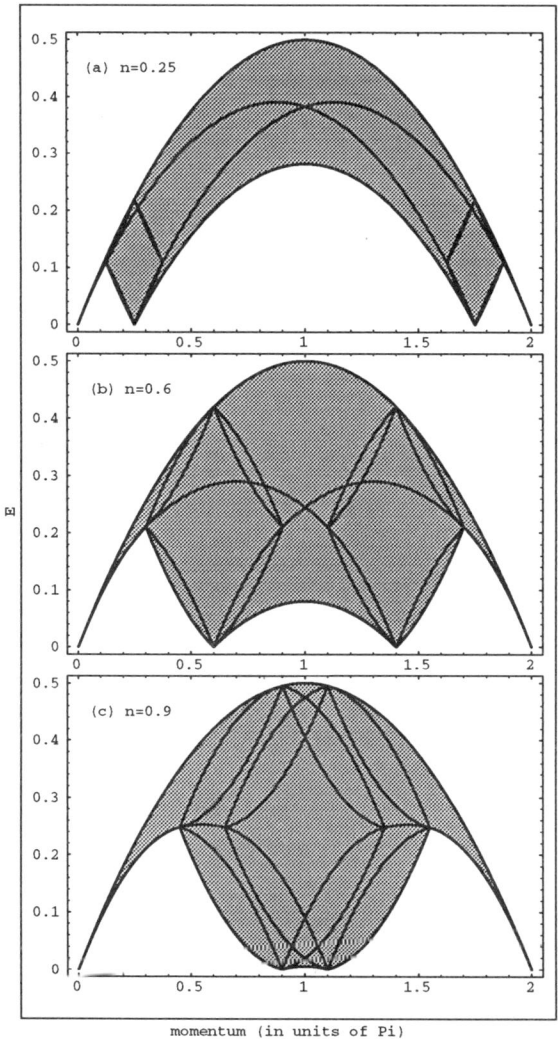

momentum (in units of Pi)

Fig. 6.3. *Compact support* for the spin-spin correlation function. $(s_L, h_L) + \bar{h} + (s_R, h_R)$, $2s_L$ and their mirror states contribute. $E = 0$ allowed at $0(2\pi)$, $2k_F$, $2\pi - 2k_F$. This indicates that $2k_F$ is the spinon Fermi point.

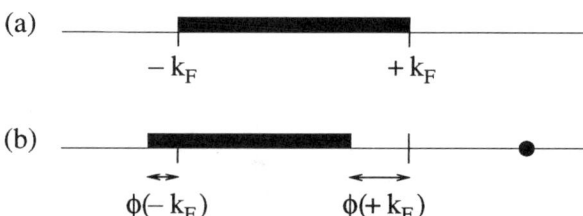

Fig. 6.4. Somewhat exaggerated phase shift at two pseudo-Fermi points as a resulting of introducing a test charge near the sea.

The "dressed charge" carried by the excitations can be calculated using the asymptotic Bethe-ansatz equations [36]. The charge enhancement of the test holon is measured by the difference in the phase shifts of the holon condensate at the pseudo-Fermi points and will in general depend on where the holon is with respect to the condensate as illustrated in Fig. 6.4. The phase shift at right pseudo-Fermi point is in general different and larger than that at left if the test charge is on the right side of the pseudo-Fermi sea. The enhanced charge is then the difference between the two phase shifts measured in units of 2π.

An exact calculation of the anti-holon charge of the hyperbolic t-J model can be carried out, and some details are given in Appendix 6.B. I find that the form of phase shift function does not explicitly depend on the parameter γ and the universal phase shift is given in the rapidity space as follow

$$\varphi(w, w^h) = \int_{-\infty}^{\infty} dv \frac{\delta_{\infty} + \text{Arctan}(2(v + w - w^h))}{\cosh(\pi v)} + \int_{-Q_0}^{Q_0} dw' \varphi(w') R(w - w'), \quad (6.11)$$

with

$$R(w) = \int_{-\infty}^{\infty} \frac{d\xi}{2\pi} \frac{e^{i\xi w}}{1 + e^{|w|}}. \quad (6.12)$$

The phase shift function is indeed independent of the form of the pseudo-momentum-rapidity map $k(w)$ which is discussed in Chapter 2. The pseudo-Fermi point Q_0 in rapidity space is determined by the density of holes and, furthermore, is in general a function of γ and fixed by the following relation

$$n_h = \int_{-Q_0}^{Q_0} \rho_0(\lambda) d\lambda, \quad (6.13)$$

where n_h is the density of holes M_h/N_a and $\rho_0(\lambda)$ is the ground state rapidity distribution function determined by the asymptotic Bethe-ansatz. When w^h resides inside the holon sea $-Q_0 < w^h < Q_0$ it corresponds to the anti-holon location. On the other hand, when $w^h > Q_0$ it corresponds to the holon excitation. The overall phase shift $\delta_{\infty} = \pm\pi/2$ is due to a background charge introduced at ∞ or $-\infty$ such that

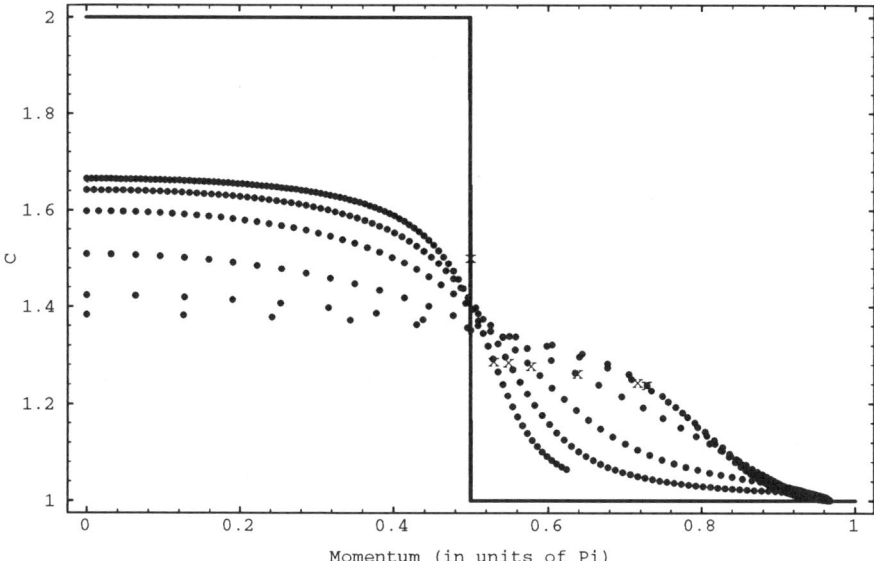

Fig. 6.5. Chare of a test (anti-)holon versus its momentum in the holon condensate for $\gamma = 0$, 0.2, 0.3, 0.5, 1.0, 2.0, ∞. Density of holes n_h is $1/2$. The pseudo-Fermi points are labeled by "x" for each γ. The charges are in units of $-e$ for the holon and $+e$ for the anti-holon where e is the electron charge and the momentum of the test holon is in units of π. (The (anti-)holons have momenta (less) greater than the pseudo-Fermi point.) The step function corresponds to the ISE model ($\gamma = 0$). The NNE model has the smallest but still considerable charge enhancement in the condensate ($\gamma = \infty$).

it is on opposite side of the sea with respect to the test charge. The phase shifts at the pseudo-Fermi points of the holon sea in the ISE limit corresponds to evaluating φ at $w = \pm\infty$ with $Q_0 \to \infty$. So, the phase shift due to the anti-holons will be independent of w^h. This function would give the anti-holon charge to be $1 + (\varphi(\infty, w^h) - \varphi(-\infty, w^h))/2\pi = 2$ independent of $|w^h| < \infty$. For the holons no charge correction is present since we must take the limit $|w^h| \to \infty$ first. Details of the calculation are presented in Appendix 6.B.

The total charge C (the bare plus the enhanced) is plotted in Fig. 6.5 as a function of the momentum of the test holon at a fixed density of electrons ($\bar{n} = 0.5$) for various values of γ. The pseudo-Fermi points of the condensate for each γ are labeled by "x". The ISE limit is given by the solid line. The curve with the smallest charge enhancement in the condensate corresponds to the NNE model. In the ISE limit, there is a clear jump in the holon charge from 1 to 2 at the pseudo-Fermi point

$\pi(1 - \bar{n})$. Therefore, if a holon is taken out the condensate, the *hole* excitation of the holon sea carries charge $+2e$ independent of where it is in the condensate. We call this hole an antiholon. For all the other values of γ, there is a considerable charge enhancement of the test holon in the condensate, and as $\gamma \to 0$ the charge approaches $-2e$.

6.6 Conclusion

In conclusion, we have devised simple rules for constructing the motifs for the excited states of the 1D ISE t-J model and identified the exact excitation content of the intermediate states for the one-particle Green's function, the charge density-density and spin-spin correlation functions. We believe that this model is in the same universality class as the NNE model, and that the most relevant states for the ground state correlation functions of the NNE model are also given by Table I. Finally, the presence of spinons, holons, and antiholons in two-dimensional models and their role in the high T_c superconductivity is an amusing possibility.

6.A   Normalization constant of the wavefunction

In this appendix the normalization constant for the ground-state wavefunction is calculated. Since it is more convenient to use the form given by Eq. (6.5), I will use it and calculate the following

$$|\Psi|^2 = \sum_{\{z_j\}} \prod_{1 \leq i < j \leq M_h} |z_i - z_j| \prod_{1 \leq i < j \leq M_h + M_\downarrow} |z_i - z_j|^{m_i m_j}, \qquad (6.A.14)$$

where $\{z_j\} \equiv \{Z_1, \ldots, Z_{M_h}, Z_1^\downarrow, \ldots, Z_{M_\downarrow}^\downarrow\}$ and $m_j = 1$ ($m_j = 2$) for the hole (down-spin) coordinate. Since the wavefunction vanishes whenever two coordinates coincide and its norm does not depend on the order of the coordinate the sum can be carried out over all configurations as follow

$$\frac{1}{M_h! M_\downarrow!} \sum_{n_1=1}^{N_a} \cdots \sum_{n_{M_h+M_\downarrow}=1}^{N_a} (-1)^{M_h(M_h-1)/2} \left(\prod_{l=1}^{M_h} z_l^{-(M_h+M_\downarrow-1)} \right) \left(\prod_{l=M_h+1}^{M_h+M_\downarrow} z_l^{-(M_h+2M_\downarrow-2)} \right)$$

$$\times \prod_{1 \leq i < j \leq M_h} (z_j - z_i) \prod_{1 \leq \alpha < \beta \leq M_h + M_\downarrow} (z_\beta - z_\alpha)^{m_\alpha m_\beta}, \qquad (6.A.15)$$

where $z_j = \exp(i 2\pi n_j / N_a)$ and the absolute value signs are removed by introducing the net currents and appropriate signs. The last two products in the sum above can be represented in Vandemonde determinants and expanded as

$$\sum_Q \sum_P \varepsilon(P) \prod_{l=1}^{M_h} z_{Q(l)}^{P(l)+l-2} \prod_{\nu=1}^{M_\downarrow} (P(M_h + 2\nu) - 1) z_{M_h+\nu}^{P(M_h+2\nu-1)-1} z_{M_h+\nu}^{P(M_h+2\nu)-2}, \qquad (6.A.16)$$

where Q and P are permutations of the sets $\{1, \ldots, M_h\}$ and $\{1, \ldots, M_h + 2M_\downarrow\}$, respectively. And, $Q(k)$ refers to the kth element of the permutation Q and $\varepsilon(P)$ is signature of the permutation P. The sum in Eq. (6.A.16) can now be expressed as

$$\frac{1}{M_h! M_\downarrow!} \sum_Q \sum_P \varepsilon(P)(-1)^{M_h(M_h-1)/2} \prod_{l=1}^{M_h} \left(\sum_{n_{Q(l)}=1}^{N_a} z_{Q(l)}^{P(l)+l-M_h-M_\downarrow-1} \right)$$

$$\times \prod_{\nu=1}^{M_\downarrow} \left(\sum_{n_{M_h+\nu}=1}^{N_a} (P(M_h + 2\nu) - 1) z_{M_h+\nu}^{P(M_h+2\nu-1)+P(M_h+3\nu)-(M_h+2M_\downarrow+1)} \right), \qquad (6.A.17)$$

where the sum over position variables n_j are taken inside the sum over permutations P and Q. Now, use following identity

$$\sum_{m=1}^{N_a} z^m = N_a \delta_{m,0} \qquad (6.A.18)$$

to reduce the first product in Eq. (6.A.17) and fix $P(l) = M_h + M_\downarrow + 1 - l$ for $l = 1, \ldots, M_h$. The sum over Q can also be performed to give $M_h!$.

The remaining $P(l)$'s are to be chosen from the set $\{1, \ldots, M_\downarrow, M_h + M_\downarrow + 1, \ldots, M_h + 2M_\downarrow\}$. The second product in Eq. (6.A.17) can now be reorganized as

$$\sum_{P'} \epsilon(P) \prod_{\nu=1}^{M_\downarrow} \sum_{n_\nu=1}^{N_a} (P(2\nu) - P(2\nu - 1)) z_\nu^{P(2\nu-1)+P(2\nu)-(M_h+2M_\downarrow+1)}, \qquad (6.A.19)$$

where the summing variable P' corresponds to only permutations that satisfy $P(2\nu) > P(2\nu - 1)$. When the sum over n_ν is performed one gets $N_a(P(2\nu) - P(2\nu - 1))$ if $P(2\nu) + P(2\nu - 1) = M_h + 2M_\downarrow + 1$ and zero otherwise. Since changing the order of a pair of indices requires that of another only the even permutations are allowed in Eq. (6.A.19) and, therefore, $\epsilon(P)$ is always equal to $+1$ in the sum. Note also the following relation

$$\sum_P \prod_{\nu=1}^{M_\downarrow} (P(2\nu) - P(2\nu - 1)) = \sum_R \prod_{\nu=1}^{M_\downarrow} (M_h + 2M_\downarrow + 1 - 2R(l))$$

$$= M_\downarrow! \prod_{j=1}^{M_\downarrow} (M_h + 2j - 1), \qquad (6.A.20)$$

where R represents the even permutations and $R(l) = P(2\nu - 1)$. Putting all the terms together one then gets the normalization constant given in Eq. (6.6) Q.E.D.

6.B Phase shift function

In this appendix the phase shift function for the hyperbolic family of integrable t-J model is calculated. The eigen-spectra are described by the following set of asymptotic Bethe-ansatz equations [96, 41]

$$N_a k(v_\alpha) = 2\pi J_\alpha + \sum_{\beta=1}^{M} \theta(v_\alpha - v_\beta) - \sum_{j=1}^{M_h} \theta(2(v_\alpha - w_j)), \qquad (6.B.21)$$

$$\sum_\alpha \theta(2(w_j - v_\alpha)) = 2\pi I_j, \qquad (6.B.22)$$

where $\theta(x) = 2\text{Arctan}(x)$ and $M = M_h + M_\downarrow$. If $M_h = 0$ the last term in the first BAE and the second BAE vanish and one gets just the form for the Heisenberg spin chain. The inverse of map $k(v)$ is also given by [55, 41]

$$v(k) = \sum_{m=1}^{\infty} \coth\left(\frac{\gamma}{2}m\right) \sin(mk) \qquad (6.B.23)$$

which needs some regularization.

One of the regularized map that is useful for γ large (i.e. the conventional Bethe-ansatz limit) is given by

$$v(k) = \frac{1}{2} \cot(k/2) + \sum_{m=1}^{\infty} \left(\coth\left(\frac{\gamma}{2}m\right) - 1\right) \sin(mk) \qquad (6.B.24)$$

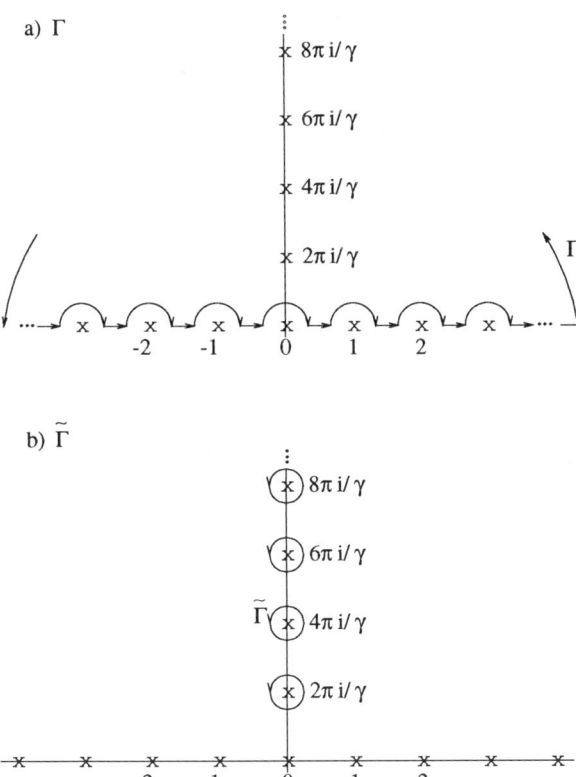

Fig. 6.6. Two loops in complex planes Γ and $\tilde{\Gamma}$ which are deformable from one to the other.

Hence, in the limit $\gamma \to \infty$ only the first term above is retained and it is identical to the map for NNE Heisenberg spin chain discussed in Chapter 2. The map above is not well behaved in the ISE limit (i.e. $\gamma \to 0$) and, thus, need to be analytically extended. The map in Eq. (6.B.23) can be represented by the following integral over Γ (see Fig. 6.6)

$$-\frac{k}{\gamma} - \int_{\Gamma_1} \frac{\coth(\gamma z/2)\sin(kz)}{1 - \exp(-2\pi i z)} + p.v. \int_{-\infty}^{\infty} dx \frac{\coth(\gamma x/2)\sin(kx)}{1 - \exp(-2\pi i x)}, \qquad (6.B.25)$$

where $p.v.$ refers to the "principal value" and the first term above is a contribution from $z = 0$ term that is needed to compensate the complex line integral. The loop Γ can be deformed to $\tilde{\Gamma}$ as shown in Fig. 6.6 and by summing up the resulting residues and doing some extra work one gets the following second expression for the map [41]

$$v(k) = -\frac{k}{\gamma} + \frac{\pi}{\gamma}\coth\left(\frac{\pi k}{\gamma}\right) - \frac{2\pi}{\gamma}\sum_{r=1}^{\infty}(f(r,k) - f(r,-k)), \qquad (6.B.26)$$

where

$$f(r,k) = \frac{1}{\exp(2\pi(2\pi r - k)/\gamma) - 1}. \qquad (6.B.27)$$

Thus, this second expression becomes $v(k) = -(k-\pi)/\gamma$ in the ISE limit (i.e. $\gamma \to 0$) where the map becomes singular.

I introduce a hole rapidity w^h and a background charge v^b at infinitely far away from the pseudo-Fermi sea. This background charge will induce a uniform phase shift and thus is dynamically decoupled from the system and, thus, I will forget this term and will just insert it at the end of the calculation. First, when w^h is not present the first BAE in the ground state where $J_{\alpha+1} - J_\alpha = 1$ can be written as

$$N_a\frac{dk(v_\alpha^0)}{kv_\alpha^0}\Delta v_\alpha^0 - \sum_\beta\frac{d\theta(v_\alpha^0 - v_\beta^0)}{dv_\alpha^0}\Delta v_\alpha^0 + \sum_j\frac{d\theta(2(v_\alpha^0 - w_j^0))}{dv_\alpha^0}\Delta v_\alpha^0 = 2\pi. \qquad (6.B.28)$$

When w^h is present there will the following shifts in the rapidities

$$v_\alpha \rightarrow v_\alpha^0 + \delta v_\alpha \qquad (6.B.29)$$
$$w_j \rightarrow w_j^0 + \delta w_j. \qquad (6.B.30)$$

Using Eq. (6.B.28) the first BAE can now be expressed to first order in δv and δw in Taylor expansion and it takes the following form

$$\varphi^{(1)}(v) = -\theta(2(v - w^h)) + \int_{-Q_0}^{Q_0}\frac{dw}{2\pi}\theta'(2(v - w))\varphi^{(2)}(w) - \int_{-\infty}^{\infty}\frac{dv'}{2\pi}\theta'(v - v')\varphi^1(v'), \qquad (6.B.31)$$

where the phase functions φ^1 and φ^2 are defined by

$$2\pi\frac{\delta v_\alpha}{\Delta v_\alpha^0} \rightarrow \varphi^{(1)}(v), \qquad (6.B.32)$$

$$2\pi\frac{\delta w_j}{\Delta w_j^0} \rightarrow \varphi^{(2)}(w). \qquad (6.B.33)$$

In other words the phase shift function is a measure of the rapidity shift with respect to the ground state rapidity spacings in units of 2π. The expression above can also be written as

$$\int_{-\infty}^{\infty}\varphi^{(1)}(v)(1 + e^{-|\xi|})e^{-i\xi v}dv = -2\int_{-\infty}^{\infty}\text{Arctan}(2(v - w^h))e^{-i\xi v}dv$$
$$+ e^{-|\xi|/2}\int_{-Q_0}^{Q_0}\varphi^{(2)}(w)e^{-i\xi w}dw \qquad (6.B.34)$$

From the second BAE one obtains the following relations

$$\varphi^{(2)}(w) = \int_{-\infty}^{\infty} \frac{dv}{2\pi} \theta'(2(w-v))\varphi^{(1)}(v).$$ (6.B.35)

Combining this with Eq. (6.B.34) one can finally obtain the following integral equation that involves only $\varphi^{(2)}(w)$

$$\varphi^{(2)}(w) = -2\int_{-\infty}^{\infty} dv \operatorname{Arctan}(2(v-w^h))D(w-v) + \int_{-Q_0}^{Q_0} dw' \varphi^{(2)}(w')R(w-w'),$$ (6.B.36)

where

$$D(x) = \int_{-\infty}^{\infty} \frac{d\xi}{2\pi} \frac{e^{i\xi x}}{2\cosh(\xi/2)},$$ (6.B.37)

$$R(x) = \int_{-\infty}^{\infty} \frac{d\xi}{2\pi} \frac{e^{i\xi x}}{1+e^{|\xi|}}.$$ (6.B.38)

By putting back the phase due do the background charge and redefining the phase function $\varphi^{(2)}(w)$ to $\varphi(w, w^h)$ one obtains the final form given in Eq. (6.11).

Chapter 7

Fractional Statistics in One-Dimension: View From An Exactly Solvable Model

In this chapter 1D fractional statistics is studied using the Calogero-Sutherland model (CSM) which describes a system of non-relativistic quantum particles interacting with inverse-square two-body potential on a ring. The inverse-square exchange can be regarded as a pure statistical interaction and this system can be mapped to an ideal gas obeying the fractional exclusion and exchange statistics. The details of exact calculations of the dynamical correlation functions for this ideal system is presented. An effective low-energy one-dimensional "anyon" model is constructed and its correlation functions are found to be in agreement with those in the CSM, and this agreement provides an evidence for the equivalence of the first- and the second-quantized construction of the 1D anyon model at least in this long wave-length limit. Furthermore, the finite-size scaling applicable to the conformally invariant systems is used to obtain the complete set of correlation exponents for the CSM.

7.1 Introduction

The fractional statistics in low-dimensional (< 3) quantum systems had been a subject appreciated only by a few [71, 73]; however, the discovery of the quantum Hall effect [68] has perhaps changed the status of the subject forever, promoting and reshaping it into one of the most profound as well as popular field in modern physics and into a necessary conceptual tool in condensed matter physics.

The fractional statistics is usually discussed in the context of two-dimensional systems where the adiabatic transport of a test particle around the others can be used to determine the statistics independent of the dynamical nature of the interacting system. In one-dimension (1D), however, the dynamics and the kinematics can not be decoupled in an unambiguous way (i.e., an exchange necessarily involves a scattering), and the assignment of the statistics to 1D particles is largely a matter of taste. On the other hand, a good taste will yield fruitful concepts and tools.

In this chapter I study the 1D fractional statistics using Calogero-Sutherland Model (CSM) [86, 11, 95] which describes a system of non-relativistic quantum particles interacting with a pairwise potential that falls off as inverse-square of the distance between the particles. One of the most important practical features of this model is that the inverse-square potential can be regarded as a pure statistical interaction and the model maps to an ideal gas of particles obeying the fractional statistics [46, 34, 35]. Usually, the first step in understanding a general class of interacting Fermi system known as Fermi liquid is to study the ideal Fermi gas which gives rise to important concepts and tools such as the Fermi surface and particle-hole excitations. Much in the same spirit I start with the simplest 1D system (i.e., the CSM) obeying the fraction statistics.

First, I recapitulate here some of the recent developments that I have already mentioned in previous chapters. A lattice cousin of the CSM called Haldane-Shastry Model (HSM), which corresponds to the $SU(2)$ Heisenberg spin chain with the inverse-square instead of the usual nearest neighbor exchange, has triggered a surge of interest in this class of models [40]. The HSM is known to possess the quantum symmetry algebra known as Yangian [42] and can be considered as a model of ideal $SU(2)$ spinon gas obeying the semionic fractional statistics [45]. The $SU(n)$ versions of the CSM [31] and the HSM [31, 67] now exist (see Chapter 4) and in particular the spectrum of $SU(n)$ HSM is known to possess the Bethe-ansatz string structure [32] (see Chapter 5). The list goes on, but I concentrate on the $U(1)$ CSM in this chapter.

The CSM is intimately related to the circular ensembles in random matrix theory first introduced by Freeman Dyson [17]. In particular, the eigenvalue distribution functions for the orthogonal, unitary, and symplectic random matrices correspond to the ground state wavefunctions of the CSM at the interaction parameters $\lambda = 1/2$, 1 and 2, respectively. ($\lambda = 1$ case corresponds to the free Fermi gas.) Some static correlation functions of the CSM can be calculated using the techniques developed for the random matrices [84].

More recently, Simons et. al. have been successful in mapping the CSM to the matrix model and to the non-linear sigma model where the supersymmetric algebra is applicable [18], and therefore are able to calculate the dynamical density-density correlation functions (DDDCF) for the CSM at $\lambda = 1/2$, 1 and 2 [93]. Haldane and Zirnbauer using the similar method calculate the one-particle Green's function at $\lambda = 2$ (i.e., the symplectic case) [114]. Forrester [21, 22] also calculates some static correlation functions at integer interaction parameters using a generalized form of the celebrated Selberg integral formula [61].

It turns out that the eigenstates of the CSM can be written in terms of Jack polynomials [22, 46] whose known algebraic properties provide a powerful and direct method for calculating the most general correlation functions. Recently, the author has been successful in calculating the exact DDDCF and one-particle Green's function at arbitrary rational interaction parameters [34, 35]. The method employed is new and is one of the main subjects in this chapter.

The exact calculations of the correlation functions further provide conclusive evidences of the inherent fractional exclusion and exchange statistics embodied in the CSM [34, 35]. I also make a direct connection of this model to the edge states of

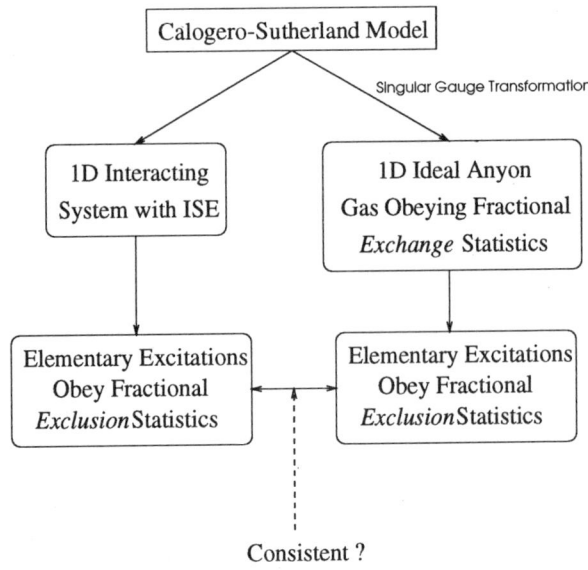

Fig. 7.1.   Logical flow of arguments leading to the consistency of the exclusion and exchange statistics.

the fractional quantum Hall droplet by first constructing an effective low-energy one-dimensional "anyon" fluid model based on a general gauge invariance argument, and then showing that the dynamical correlation functions agree with those of the CSM. This connection between the CSM and the edge states has previously been suggested by various people [48].

This chapter is organized in the following way. In Section 7.2 I show how to construct general eigenstates of the CSM, and in Section 7.3 some properties of the Jack polynomials are introduced. In Section 7.4 some key aspects of the exclusion and the exchange statistics are discussed. I show how to calculate the DDDCF and the one-particle Green's function in Section 7.5 and 7.6. By examining the excitation contents of the intermediate states for the correlation function I conclude that the exclusion and the exchange statistics are consistent as was first shown by the author in [34]. Fig. 7.1 shows the logical structure of this consistency argument. As will be shown later in this chapter there are two ways to look at CSM: (i) as an interacting system of impenetrable particles where the quantum exchange statistics is decoupled or (ii) as an ideal gas of particles obeying fractional exchange statistics. The elementary excitations in either case are shown to satisfy the *same* exclusion statistics. This consistency gives one confidence to consider the CSM as an ideal gas of 1D anyons.

In Section 7.7 the "harmonic-fluid" description of the anyon fluid is constructed and the dynamical correlation functions calculated; and I further show that they agree with those of the CSM. This agreement provides an explicit connection between the CSM and the system of coupled left- and right-edges of the fractional quantum Hall effect and further shows the equivalence between the first- and the second-quantized construction of one-dimensional "anyon" gas at least in the long-wavelength limit. In Section 7.8 the finite-size scaling as applied to the conformally invariant systems is used to obtain the complete set of correlation exponents. I also discuss some aspect of the lattice cousins of the CSM in Section 7.9.

7.2 Eigenstates of Calogero-Sutherland Model

In this section I introduce the CSM and show how to construct the general eigenstates of the model following Sutherland [95]. First, the Hamiltonian for the CSM on a ring of length L is given by

$$H = -\sum_{j=1}^{N} \frac{\partial^2}{\partial x_j^2} + \sum_{j<l} \frac{2\lambda(\lambda - 1)}{d^2(x_j - x_l)}, \tag{7.1}$$

where $\hbar^2/2m = 1$ and $d(x) = (L/2\pi)|\sin(\pi x/L)|$. The ground state wavefunctions of the model at integer values of λ corresponds to 1D versions of Laughlin's wavefunctions [72] and are given by

$$\Psi_0 = \prod_{i<j}(z_i - z_j)^\lambda \prod_k z_k^{J_0}, \tag{7.2}$$

where the current $J_0 = -\lambda(N-1)/2$ and $z_j = \exp(i2\pi x_j/L)$. When $0 < \lambda < 1$ there is another possible ground state with power $1 - \lambda$; however, only the solution with power λ will be considered for reasons of continuity with $\lambda > 1$ solutions.

The excited states of this model are constructed by multiplying some symmetric polynomials to the ground state wavefunction, and this construction is analogous to that of the gapless edge excitations of the quantum Hall effect [104]. A general excited state $\Psi_n^\lambda = \Psi_0 J_n^\lambda$ is labeled by the quantum numbers $\mathbf{n} = (n_1, n_2, \ldots, n_N)$, and J_n^λ satisfies the following new eigenvalue equation

$$\tilde{H} J_n^\lambda = E_n J_n^\lambda, \tag{7.3}$$

where $\tilde{H} = H_0 + \lambda H_1$, and

$$H_0 = \sum_{j=1}^{N}(z_j \partial_{z_j})^2, \tag{7.4}$$

$$H_1 = \sum_{j<k} \frac{z_j + z_k}{z_j - z_k}(z_j \partial_{z_j} - z_k \partial_{z_k}). \tag{7.5}$$

The eigenstates of the new Hamiltonian $\tilde{H}$ are represented in terms of the following bosonic basis states

$$\Phi(\mathbf{n}) = \sum_P \prod_{j=1}^N z_j^{n_{P_j}}, \tag{7.6}$$

where the sum extends over all permutations of the integer set $\mathbf{n}$ which can be considered as a set of bosonic quantum numbers with no restrictions on their values. Since $\Phi(\mathbf{n})$ does not depend on the ordering of the quantum numbers, let $n_1 \geq n_2 \geq \ldots \geq n_N$ without loss of generality. These symmetric polynomials form a complete basis.

The action of $\tilde{H}$ on $\Phi(\mathbf{n})$ can be easily calculated and are given by

$$H_0\Phi(\mathbf{n}) = \left(\sum_{j=1}^N n_j^2\right)\Phi(\mathbf{n}), \tag{7.7}$$

$$H_1\Phi(\mathbf{n}) = \sum_{j<k}(n_j - n_k)\left(\Phi(\mathbf{n}) + 2\sum_{s=1}^{n_j-n_k-1}\Phi(\ldots, n_j - s, \ldots, n_k + s, \ldots)\right) \tag{7.8}$$

H_0 generates only the basis state $\Phi(\mathbf{n})$ itself while H_1 is responsible for generating a family of states $\Phi(\ldots, n_j - s, \ldots, n_k + s, \ldots)$ which are obtained from $\Phi(\mathbf{n})$ by all possible pairwise "squeezing" of the quantum numbers. If a state is generated from another by squeezing a pair of quantum numbers by one unit (i.e., $n_j \to n_j - 1$, $n_k \to n_k + 1$ for $n_j - n_k \geq 2$), then I call the former "daughter-state" and the latter its "mother-state."

The family of states can be organized into *levels* such that the members of a given level are mutually not related or *unreachable* (i.e., no mother-daughter relationship exists between the members in the same level) and the daughters of a member from a given level always belong to a *lower* level in the family. The highest-level mother-state denoted by $|\nu\rangle_1$, where ν is a level index and is equal to the total number of levels in the family, generates the entire family of daughter-states which are denoted by $|\mu\rangle_\alpha$ where $1 \leq \mu < \nu$ and α an index for the states in the μth level. While the order of the levels in the family can be uniquely determined, the order within a level is quite arbitrary.

To provide an illustration of the above mentioned family structure, I give the following example. Let the highest-level mother-state be $|6\rangle_1 = \Phi(4,3,1,0)$. Then, the members of the family are $|5\rangle_1 = \Phi(4,2,2,0)$, $|4\rangle_1 = \Phi(4,2,1,1)$, $|4\rangle_2 = \Phi(3,3,2,0)$, $|3\rangle_1 = \Phi(3,3,1,1)$, $|2\rangle_1 = \Phi(3,2,2,1)$, and $|1\rangle_1 = \Phi(2,2,2,2)$. A pictorial representation of the family structure is shown in Fig. 7.2 and I call it a level diagram. In Fig. 7.2 the states are represented by dots and each arrow connects a state A to B where B is *reachable* from A by *squeezing* on a pair of quantum numbers for A by one unit (i.e., connects a mother and her daughter). The levels are ordered from top to bottom, from the most squeezable to unsqueezable states, such that the highest-level mother-state is at the top; and therefore the arrows always point downwards and never upwards. A set of arrows that is topologically equivalent to a directed

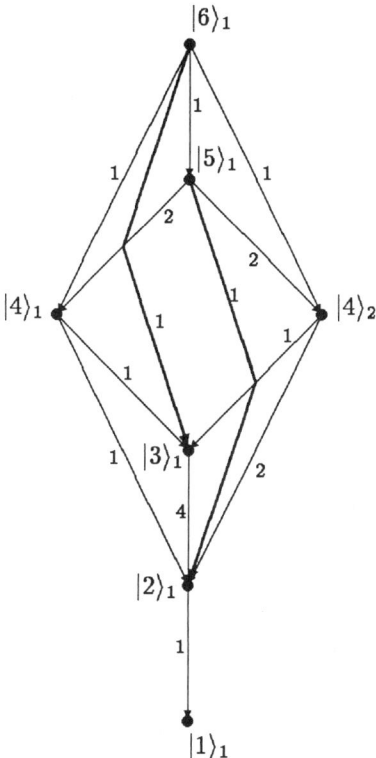

Fig. 7.2. A level diagram for a family with the highest-level mother-state given by $|6\rangle_1 = \Phi(4,3,1,0)$. The arrows connect the mother states to their daughter states that are generated by squeezing on the pairs of quantum numbers by one unit.

line forms a *path*. A *weight* W of an arrow is defined to be $m(n_i)m(n_j)$, where n_i and n_j are the two quantum numbers squeezed to produce a daughter and $m(l)$ the multiplicity of l in the quantum number set specifying the mother-state. In Fig. 7.2 the numbers next to the arrows are the corresponding weights. The states in the same level are not connected; hence, the Hamiltonian is diagonal in that subspace.

A *subfamily* of a family can also be constructed by starting from a given state, which would be the highest-level mother-state of that subfamily, and grouping all her *reachable* off-springs. The total number of *subfamilies* is therefore equal to the *dimension* (i.e., the total number of basis states) of the family.

The lowest-level states (the unsqueezable states) clearly have to be of one of the two types: (I) $\Phi(\ldots, m, m, m, m, m, \ldots)$ or (II) $\Phi(\ldots, m, m, m, m, m-1, m-1, m-1, m-1, \ldots)$. The type-I state corresponds to the ground state up to a global Galilean boost and the type-II to states with a single hole excitation. These one-hole states are eigenstates of $\tilde{H}$ and, furthermore, since the following superposed state

$$\Psi(x) = \prod_{j=1}^{N}(z - z_j), \tag{7.9}$$

where $z = \exp(i2\pi x/L)$, can be expanded purely in terms of the one-hole eigenstates, $\Psi(x)$ describes a state with a hole localized at x.

The matrix representation of $\tilde{H}$ acting on the partially ordered state space is always triangular since the action of $\tilde{H}$ on a given state always generates states belonging to lower levels. The eigenvalues, therefore, are simply given by the diagonal matrix elements. In particular the energy of an eigenstate spanned by a family with the highest-level mother-state $\Phi(\mathbf{n})$ is given by

$$E_{\mathbf{n}}^{(0)} = \sum_{j=1}^{N} n_j^2 + \lambda \sum_{j<k}(n_j - n_k). \tag{7.10}$$

The off-diagonal elements are given by

$$_\alpha\langle\mu|\tilde{H}|\nu\rangle_\beta = \begin{cases} \left[\sum_{\mathcal{P}}\left(\prod_{i\in\mathcal{P}} W_i\right)\right] E_{\mathbf{n}}^{(1)} & \text{if } \nu > \mu, \\ 0 & \text{if } \nu \le \mu, \end{cases} \tag{7.11}$$

where $E_{\mathbf{n}}^{(1)} = 2\lambda\sum_{j<k}(n_j - n_k)$, and the sum is over all possible path $\mathcal{P}$ from $|\nu\rangle_\beta$ to $|\mu\rangle_\alpha$ and the product over all the weights W_i of the intermediate arrows belonging to $\mathcal{P}$.

For the example given in Fig. 7.2, $\tilde{H}$ is represented by

$$
\tilde{H} =
\begin{pmatrix}
\varepsilon_{(1,1)}^{(1,1)} & \varepsilon_{(2,1)}^{(1,1)} & \varepsilon_{(3,1)}^{(1,1)} & \varepsilon_{(4,1)}^{(1,1)} & \varepsilon_{(4,2)}^{(1,1)} & \varepsilon_{(5,1)}^{(1,1)} & \varepsilon_{(6,1)}^{(1,1)} \\
0 & \varepsilon_{(2,1)}^{(2,1)} & \varepsilon_{(3,1)}^{(2,1)} & \varepsilon_{(4,1)}^{(2,1)} & \varepsilon_{(4,2)}^{(2,1)} & \varepsilon_{(5,1)}^{(2,1)} & \varepsilon_{(6,1)}^{(2,1)} \\
0 & 0 & \varepsilon_{(3,1)}^{(3,1)} & \varepsilon_{(4,1)}^{(3,1)} & \varepsilon_{(4,2)}^{(3,1)} & \varepsilon_{(5,1)}^{(3,1)} & \varepsilon_{(6,1)}^{(3,1)} \\
0 & 0 & 0 & \varepsilon_{(4,1)}^{(4,1)} & 0 & \varepsilon_{(5,1)}^{(4,1)} & \varepsilon_{(6,1)}^{(4,1)} \\
0 & 0 & 0 & 0 & \varepsilon_{(4,2)}^{(4,2)} & \varepsilon_{(5,1)}^{(4,2)} & \varepsilon_{(6,1)}^{(4,2)} \\
0 & 0 & 0 & 0 & 0 & \varepsilon_{(5,1)}^{(5,1)} & \varepsilon_{(6,1)}^{(5,1)} \\
0 & 0 & 0 & 0 & 0 & 0 & \varepsilon_{(6,1)}^{(6,1)}
\end{pmatrix} ,
\tag{7.12}
$$

where $\varepsilon_{(\nu,\beta)}^{(\mu,\alpha)} = {}_\alpha\langle\mu|\tilde{H}|\nu\rangle_\beta$. For example, $\varepsilon_{(6,1)}^{(3,1)} = 7 \times E_{(3,3,1,1)}^{(1)}$ since there are five different ways to get from $|6\rangle_1$ to $|3\rangle_1$ with the following corresponding weights (see Fig. 7.2):

1. $|6\rangle_1 \xrightarrow{1} |4\rangle_1 \xrightarrow{1} |3\rangle_1$;

2. $|6\rangle_1 \xrightarrow{1} |5\rangle_1 \xrightarrow{2} |4\rangle_1 \xrightarrow{1} |3\rangle_1$;

3. $|6\rangle_1 \xrightarrow{1} |3\rangle_1$;

4. $|6\rangle_1 \xrightarrow{1} |4\rangle_2 \xrightarrow{1} |3\rangle_1$;

5. $|6\rangle_1 \xrightarrow{1} |5\rangle_1 \xrightarrow{2} |4\rangle_2 \xrightarrow{1} |3\rangle_1$.

The eigenenergy given by Eq. (7.10) plus the ground state energy can be rewritten in terms of newly defined pseudo-momenta k_j as

$$
E = \frac{\hbar^2}{2m} \sum_{j=1}^{N} k_j^2,
\tag{7.13}
$$

where

$$
Lk_j = 2\pi I_j + \pi(\lambda - 1) \sum_{l-1}^{N} \operatorname{sgn}(k_j - k_l).
\tag{7.14}
$$

The quantum numbers I_j are now distinct (half-odd) integers and are related to n_j's by $I_j = n_j + (N + 1 - 2j)/2$.

The distribution of k_j determined by Eq. (7.14) is used to construct pictorial representations of the eigenstates called motifs which are crucial for exposing the fractional statistics obeyed by the elementary excitations of the model. Detailed discussion of this subject is given in Section 7.4.

7.3   Jack Symmetric Polynomials

The polynomial solutions of Eq. (7.3) in Section 7.2 is also known in mathematical literature as Jack polynomials [56]. In fact, Stanley [94] has shown that the complete set of linearly independent solutions of Eq. (7.3) is indeed given by Jack polynomials up to global Galilean boosts (i.e., up to the factor $\prod_{j=1}^{N} z_j^J$ where J is the current and takes an arbitrary real number) [34, 35].

For better readability this section is divided into two subsections: the first introduces the conventional notations used in mathematical literatures and the second some of the general properties of Jack polynomials.

7.3.1   Introduction to notations

Partitions are defined as sequences of non-negative integers in non-increasing order and are used to label the symmetric polynomials. They are denoted by bold-face Greek letters as

$$\kappa = (\kappa_1, \kappa_2, \ldots, \kappa_N), \tag{7.15}$$

where $\kappa_1 \geq \kappa_2 \geq \ldots \geq \kappa_N$. Non-zero κ_j are called *parts* of κ whose *length* (i.e., the total number of non-zero *parts*) is denoted by $\ell(\kappa)$. The *weight* of the partition is defined by $|\kappa| = \sum_{j=1}^{\ell(\kappa)} \kappa_j$. If $\kappa_1 + \ldots + \kappa_i \geq \mu_1 + \ldots + \mu_i$ for all $i \geq 1$, then $\kappa \geq \mu$.

Young diagram $\mathcal{D}(\kappa)$ is used to graphically represent a partition: $\mathcal{D}(\kappa) = \{(i,j) : 1 \leq i \leq \ell(\kappa), 1 \leq j \leq \kappa_i\}$. The cell labeled by (i,j) is situated in the i-th row and the j-th column of the Young diagram. The diagram of κ, therefore, consists of $\ell(\kappa)$ rows of lengths κ_j.

A *conjugate* of κ is denoted by $\kappa' = (\kappa_1', \kappa_2', \ldots)$ and corresponds to a partition whose diagram is obtained by changing all the rows of $\mathcal{D}(\kappa)$ to columns in non-increasing order from the left to right. For example, the conjugate of $\kappa = (5,2,2,1)$ is $\kappa' = (4,3,1,1,1)$. Now, the following simple but useful identity can be derived [82]

$$n(\kappa) \equiv \sum_{i=1}^{\ell(\kappa)} (i-1)\kappa_i = \sum_{i=1}^{\ell(\kappa')} \binom{\kappa_i'}{2}. \tag{7.16}$$

In order prove Eq. (7.16) every cell in the ith row of $\mathcal{D}(\kappa)$ is filled in with an integer $i-1$. Since $n(\kappa)$ corresponds to the sum of all the integers in the diagram, the two different expressions for $n(\kappa)$ are obtained depending on whether the numbers in each row or column are summed first.

For a given cell $s = (i,j)$ of a diagram $\mathcal{D}(\kappa)$ there are corresponding *arm-length* $a(s) = \kappa_i - j$, *arm-colength* $a'(s) = j - 1$, *leg-length* $l(s) = \kappa_j' - i$, and *leg-colength* $l'(s) = i - 1$. The *upper* and *lower* hook-lengths are defined, respectively, as

$$h_\kappa^*(s) = l(s) + \frac{1 + a(s)}{\lambda}, \tag{7.17}$$

$$h_*^\kappa(s) = l(s) + 1 + \frac{a(s)}{\lambda}. \tag{7.18}$$

Refer to Fig. 7.3 for an illustration.

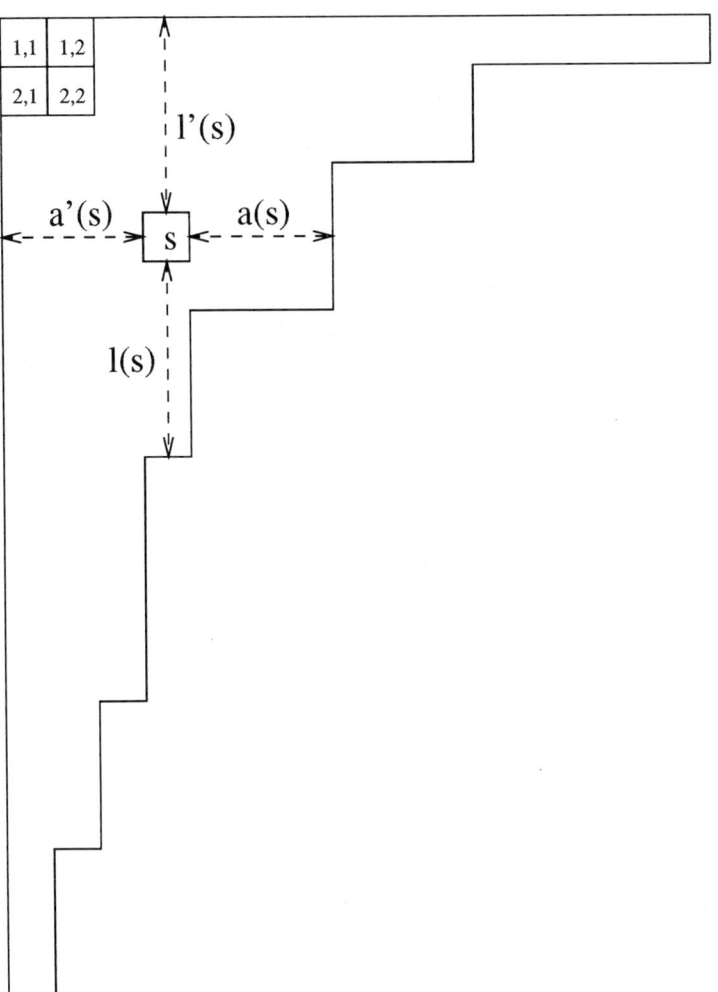

Fig. 7.3. Illustration of arm-(co)lengths $a(s)(a'(s))$ and leg-(co)lengths $l(s)(l'(s))$ for a partition.

7.3.2 General properties of Jack polynomials

The symmetric polynomials are indexed by the partitions. The bosonic basis functions $\Phi(\mathbf{n})$ of the CSM are called the monomial symmetric functions and the quantum numbers $\mathbf{n}$ correspond to the partitions defined in the previous subsection. Since the quantum numbers are allowed to be negative integers, the correspondence is only up to some trivial constant translation or the global Galilean boost. This restriction to non-negative integer parts is more of a convenience than a restriction since the CSM Hamiltonian is invariant under the global Galilean transformation.

I shall denote Jack symmetric polynomials as $J_\kappa^{1/\lambda}(z_1, \ldots, z_N)$ which are the solutions of Eq. (7.3). If $\lambda = 1$, Jack polynomials reduce to Shur functions which describe the excited states of the free fermions. At $\lambda = 0$, it becomes the monomial symmetric function which is just the free bosonic wavefunction. If $\lambda = 2$ or $1/2$, they are called the zonal spherical functions. As $\lambda \to \infty$, $J_\kappa^{1/\lambda}$ reduce to the elementary symmetric functions.

One way of defining Jack polynomials is through the differential equation (7.3). The other is based on the properties of the power-sum symmetric function $p_\kappa = p_{\kappa_1} p_{\kappa_2} p_{\kappa_3} \cdots$, where $p_{\kappa_\nu} = \sum_j z_j^{\kappa_\nu}$. Define a bilinear scalar product on the vector space of all symmetric functions of finite degree as

$$\langle p_\kappa, p_\mu \rangle_{1/\lambda} = \delta_{\kappa,\mu} z_\kappa \lambda^{-\ell(\kappa)}, \tag{7.19}$$

where $z_\kappa = \prod_{i \geq 1} i^{m_i} m_i!$, and $m_i = m_i(\kappa)$ is the number of parts of κ equal to i. Using this definition, Macdonald [82] proved that there are unique symmetric functions satisfying the following three properties:

1. *Orthogonality*: $\langle J_\kappa, J_\mu \rangle_{1/\lambda} = \delta_{\kappa,\mu} j_\kappa^\lambda$, where j_κ^λ is the normalization constant.

2. *Triangularity*: $J_\kappa = \sum_\mu v_{\kappa\mu} \Phi(\mu)$, where $v_{\kappa\mu} = 0$ unless $\kappa \leq \mu$.

3. *Normalization*: If $|\kappa| = n$, then $v_{\kappa\mu} = n!$, where $\kappa = \underbrace{(1, 1, \ldots, 1)}_{n}$.

J_κ are, then, constructed by Gram-Schmidt orthogonalization relative to the scalar product on the ring of polynomials. Stanley [94] proved that the normalization constant is given by

$$j_\kappa^\lambda = \prod_{s \in \kappa} h_\kappa^*(s) h_*^\kappa(s). \tag{7.20}$$

There is another scalar product on which Jack polynomials are orthogonal:

$$
\begin{aligned}
\langle \kappa | \mu \rangle_{1/\lambda} &\equiv A_N^2 \left(\prod_{j=1}^N \int_0^L dx_j \right) \overline{J_\kappa^{1/\lambda}(z_1, z_2, \ldots, z_N)} J_\mu^{1/\lambda}(z_1, z_2, \ldots, z_N) \prod_{i<j} |z_i - z_j|^{2\lambda} \\
&= A_N^2 j_\kappa^\lambda \prod_{s \in \kappa} \frac{N + a'(s)/\lambda - l'(s)}{N + (a'(s) + 1)/\lambda - (l'(s) + 1)} \delta_{\kappa,\mu},
\end{aligned} \tag{7.21}
$$

where $z_j = \exp(i2\pi x_j/L)$ and $A_N^2 = (1/L)^N \Gamma^N(1+\lambda)/\Gamma(1+\lambda N)$, and the bar over the polynomial denotes the complex conjugation. Note also that the multidimensional integral above is equal to L^N times the constant term in

$$J_\kappa^{1/\lambda}(1/z_1, 1/z_2, \ldots, 1/z_N) J_\mu^{1/\lambda}(z_1, z_2, \ldots, z_N) \prod_{i \neq j}(1 - \frac{z_i}{z_j})^\lambda. \qquad (7.22)$$

Eq. (7.21) has been conjectured first by Macdonald [81] and then later proved by himself [82] and also by Kadell [59].

Since Jack polynomials span the vector space of symmetric functions, they can be used to expand any symmetric functions. This property is particularly useful in calculating the correlation functions of the CSM as will be shown later in this chapter. Here are some of them [51, 108]:

$$\sum_{i=1}^{N} z_i^n = \frac{n}{\lambda} \sum_{|\kappa|=n} \frac{[0']_\kappa^\lambda}{j_\kappa^\lambda} J_\kappa^{1/\lambda}(z_1, \ldots, z_N), \qquad (7.23)$$

$$\prod_{j=1}^{N}(1-z_j)^a = \sum_\kappa \frac{\{-a\}_\kappa^\lambda}{\lambda^{|\kappa|} j_\kappa^\lambda} J_\kappa^{1/\lambda}(z_1, \ldots, z_N), \qquad (7.24)$$

where $[a]_\kappa^\lambda = \prod_{(i,j)\in\kappa}\{a+(j-1)/\lambda-(i-1)\}$, and $\{a\}_\kappa^\lambda = \prod_{(i,j)\in\kappa}\{a-\lambda(i-1)+(j-1)\}$. The sum in Eq. (7.24) extends over all possible partitions while in Eq. (7.23) it is restricted to partitions with *weight* $|\kappa| = n$. The prime in $[0']_\kappa^\lambda$ denotes that the product does not include the cell $(0,0)$; otherwise the total product is trivially equal to zero. J_κ also satisfies

$$J_\kappa^{1/\lambda}(wz_1, wz_2, \ldots, wz_N) = w^{|\kappa|} J_\kappa^{1/\lambda}(z_1, z_2, \ldots, z_N), \qquad (7.25)$$

since Jack polynomials are homogeneous functions of degree $|\kappa|$.

7.4   Fractional statistics

I divide this section into two subsections. In the first (second) subsection the fractional *exchange* (*exclusion*) statistics is discussed in the context of one dimensional models. The exchange statistics in two-dimension is directly relevant to the fractional quantum Hall effect and various people have made contributions to this fascinating field [90]. In one-dimension, however, the definition of fractional exchange statistics is rather obscure and incomplete with a possible exception of the CSM. One the other hand, the definition of fractional exclusion statistics is spatial dimension independent and is based on the structure of the Hilbert space rather than the configuration space. While the fermions obey the well-known Pauli exclusion principle, more exotic particles may obey a "generalized exclusion principle." [44]

7.4.1 Exchange Statistics

The first full mathematical treatment of the fractional statistics is due mainly to Leinaas and Myrheim who used the multiply connected topological structure of the configuration space of collections of identical particles to show the possibility of exotic statistics in spatial dimension less than three [73]. Fig. 7.4 illustrates how two-dimension is qualitatively different from three-dimension by showing respective two-particle configuration spaces.[1] Note that in 3D two-exchanges represented by the big circle on the sphere can be continuously shrunken to a point. Hence, two-exchange is equivalent to no exchange up to two possible phases +1 and −1 which correspond to bosons and fermions, respectively. In 2D, however, there are infinitely many topologically distinct exchanges and, therefore, there are no restrictions for the allowed exchange phases and arbitrary statistics are possible. While this idea has been extensively applied to two-dimensional systems especially in the context of the fractional quantum Hall effect, very little attention has been paid to the one-dimensional (1D) systems. Perhaps the main difficulty in 1D systems is that the physical exchanges of particles necessarily involve scattering processes and that there is no known unique way of un-tangling the kinematic aspect of the fractional statistics from the dynamical processes.

For integrable 1D quantum systems, however, there is a general consistency condition known as the Yang-Baxter equation (YBE) which essentially puts strong restrictions on the scattering matrices. Therefore, intuitively the braiding of the particle "trajectories" in one-dimension when properly defined may be given by the YBE [24]. In fact the YBE is known to be intimately related to the the the knot theory and the braid groups [57]. Hence, perhaps the following quantum Yang-Baxter equation should be interpreted as the one-dimensional generalized braiding relations,

$$S_{12}(v - u)S_{13}(v)S_{23}(u) = S_{23}(u)S_{13}(v)S_{12}(v - u), \qquad (7.26)$$

where $S_{ij}(v)$ is the scattering matrix in the tensor product of linear vector spaces, $V \otimes V \otimes V$ and acts non-trivially only in the ith and jth space, e.g. $S_{12}(v) = R(v) \otimes I$ where $R(v)$ is a matrix defined in $V \otimes V$. The parameters v and u are called spectral parameters and are equal to the rapidity differences between two colliding particles. One can rewrite Eq. (7.26) in terms of $\tilde{R}(v) = PR(v)$ where P denotes the transposition, $Px \otimes y = y \otimes x$. Define matrices $\tilde{R}_i(v) = I \otimes \cdots \otimes \tilde{R}(v) \otimes \cdots \otimes I$ on $V \otimes \cdots \otimes V$, where $\tilde{R}(v)$ acts on i-th and $i + 1$-th spaces. The matrices $\tilde{R}_i(v)$ satisfy the following relations

$$\tilde{R}_i(v)\tilde{R}_j(u) = \tilde{R}_j(u)\tilde{R}_i(v) \quad \text{if } |i - j| \geq 2, \qquad (7.27)$$

$$\tilde{R}_{i+1}(v - u)\tilde{R}_i(v)\tilde{R}_{i+1}(u) = \tilde{R}_i(u)\tilde{R}_{i+1}(v)\tilde{R}_i(v - u). \qquad (7.28)$$

Without the spectral parameters the $\tilde{R}$ matrices satisfy the braiding relations. The

---

[1]The particles need to be impenetrable so that the locus of positions of one particle with respect to the other is topologically equivalent to a sphere for 3D and a circle for 2D.

a) 3D

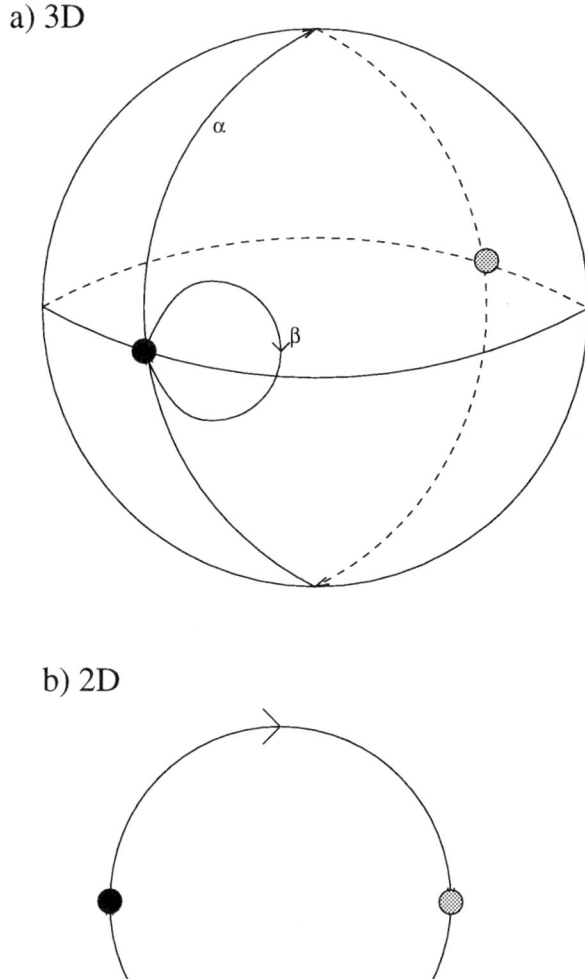

b) 2D

Fig. 7.4. Configuration spaces for a pair of impenetrable particles in three- and two-dimensions. While there only two topologically distinct exchanges in three-dimension there are infinitely many in two-dimension.

complete description of the fractional statistics based on the YBE needs to be worked out.

At present it seems unclear to me how the construction based on the YBE can be applied to a general 1D (i.e., not necessarily integrable) systems. Maybe there is a way to map a class of general 1D models to integrable models with in some sense small residual interactions that break the integrability. Instead of pursuing this incomplete Yang-Baxter story any further, I give here somewhat heuristic but intuitive device for constructing the 1D Abelian fractional statistics. The N-particle state is constructed as follow

$$\int dx_1 \cdots dx_N \Psi(x_1, \ldots, x_N, t) \phi^*(x_N, t) \cdots \phi^*(x_1, t)|0\rangle, \qquad (7.29)$$

where $\phi^*(x, t)$ is the anyon creation operator. The "anyon" fields satisfy the following equal time commutation relations

$$[\phi(x,t), \phi(y,t)]_\lambda = [\phi^*(x,t), \phi^*(y,t)]_\lambda = [\phi(x,t), \phi^*(y,t)]_\lambda = 0 \text{ if } x \neq y, \qquad (7.30)$$

where $[x, y]_\lambda = xy + \exp(i\pi\lambda \text{sgn}(y - x))yx$.

The wavefunction $\Psi(x_1, \ldots, x_N, t)$ is multi-valued and satisfies

$$\Psi(\ldots, x_i, \ldots, x_j \ldots, t) = e^{i\pi\lambda \text{sgn}(x_i - x_j)}\Psi(\ldots, x_j, \ldots, x_i, \ldots, t). \qquad (7.31)$$

The integrand in Eq. (7.29), however, is always single valued since the phases arising from the wavefunction and the "anyon" field operators are set to cancel each other.

If I specialize to the $U(1)$ CSM the fractional exchange statistics can be formulated in the first quantized language. First, the Hamiltonian (7.1) can be rewritten as [89]

$$H = \frac{1}{2m} \sum_{j=1}^{N} \left(p_j + i\frac{\pi\hbar\lambda}{L} \sum_{k(\neq j)} \cot\left[\frac{\pi(x_j - x_k)}{L}\right] P_{jk} \right)^2, \qquad (7.32)$$

where $p_j = -i\hbar\partial_{x_j}$ is the momentum operator and P_{ij} the particle exchange operator. The extra term added to the momentum operator is an 1D analog of the Chern-Simons gauge field.

In two-dimension there are two well-known ways to code the fractional statistics for the ideal anyon gas. One is to take the free Hamiltonian and require that its wavefunctions be multi-valued as in Eq. (7.31). The other is to introduce the Chern-Simons gauge field and write the Hamiltonian in terms of this gauge field and further require that the wavefunctions be single valued and symmetric. The Hamiltonian (7.32) corresponds to the 1D version of the second formulation of the fractional exchange statistics. Hence, the following symmetric wavefunctions are the eigenstates of the 1D anyon system

$$\Psi_\kappa^\lambda(x_1, \ldots, x_N) = \varphi(x_1, \ldots, x_N)\Psi_0 J_\kappa^\lambda, \qquad (7.33)$$

where the "ordering function" φ is introduced to maintain the total wavefunction symmetric [34, 35]. In particular φ keeps track of the braiding of the particles and is set to cancel the exchange phases arising from Ψ_0. I require the ordering function to satisfy

$$\varphi(x_1, \ldots, x_m, \ldots, x_N) = e^{i(m-1)\pi\lambda}\varphi(x_1, \ldots, \hat{x}_m, \ldots, x_N), \qquad (7.34)$$

where the hat over the variable denotes the absence of that variable from the function. In other words, to remove the particle at x_m it is necessary to pass through $m - 1$ particles.

In Section 7.7 I construct explicit second quantized "anyon" fields and show their consistency with the first quantized formulation of the fractional exchange statistics for the CSM.

7.4.2 Exclusion Statistics

The notion of fractional *exclusion* statistics based on the so called "generalized Pauli exclusion principle" has first been formulated by Haldane and applied to the elementary topological excitations of general condensed matter systems [44]. This new concept of statistics is based on the structure of the single particle Hilbert space of the elementary excitations. More specifically, the change in the size of the available states (ΔD) in the Hilbert space as the number of particles (i.e., the elementary excitations) is changed (ΔN) for a given system with fixed boundary condition defines the statistics of the particles with the statistical parameter defined as $g = -\Delta D/\Delta N$. Hence, for example, the bosons and fermions are identified with $g = 0$ and $g = 1$, respectively.

In order to facilitate proper understanding of the *exclusion* statistics in the context of the CSM I introduce a pictorial representation of the eigenstates and make the identification of the excitation contents of the states easier [34, 35]. Eq. (7.14) gives the occupation configurations of the pseudo-momenta k_j for all the eigenstates of the CSM. The quantum numbers $\{I_j\}$ in Eq. (7.14) are distinct (half-odd) integers and in the ground state are given by the following set

$$\{I_j^0\} = \left\{-\frac{N-1}{2}, -\frac{N-3}{2}, \ldots, \frac{N-3}{2}, \frac{N-1}{2}\right\}. \qquad (7.35)$$

Therefore, the ground state pseudo-momenta are given by

$$\{k_j^0\} = \left\{-\frac{\pi\lambda}{L}(N-1), -\frac{\pi\lambda}{L}(N-3), \ldots, \frac{\pi\lambda}{L}(N-3), \frac{\pi\lambda}{L}(N-1)\right\}. \qquad (7.36)$$

The total ground state energy $E^0 = \sum_j (k_j^0)^2$ is equal to $\pi^2\lambda^2 N(N^2 - 1)/3L^2$. The excited states are given by integer displacements of $\{I_j^0\}$. Therefore, two neighboring pseudo-momenta for any arbitrary state must be separated by

$$\Delta k_j \equiv |k_j - k_{j-1}| = \frac{2\pi}{L}(\lambda + l), \qquad (7.37)$$

where l is a non-negative integer.

In order to construct a picture that exposes the excitation content of the excited states, I let λ to be a rational number p/q with p and q coprimes and introduce one-dimensional lattice with the lattice spacing equal to $2\pi/qL$. I assign each lattice point with 1 if that lattice point coincides with the value of one of the occupied pseudo-momenta and with 0 if it does not. Hence, the ground state for $\lambda = 3/2$ and $N = 10$ is represented by $\ldots 00000000100100100100100100100100100100001000000\ldots$. All the other excited states can be obtained from this ground state configuration by displacing the *ones* such that the number of *zeroes* between any pair of *ones* is equal to $p - 1 + ql$ where l is a non-negative integer.

I use three different names for the particles in the model—real, pseudo, and quasi-particles. The real particles are, of course, the physical quantum particles described by the canonically conjugate coordinate and momentum variables $\{x_j, p_j\}$. The pseudo-particles are described by the pseudo-momentum operator (see Eq. (7.32)) which a sum of the usual momentum and the 1D "statistical gauge field." The quasiparticles are the elementary excitations of the system. Because the pseudo-particles form an ideal gas, the quasiparticles are essentially same as the pseudo-particles excited out of the condensate. The holes left behind in the pseudo-particle condensate will be called quasiholes. Hence, the name "pseudo" and "quasi" will be used interchangeably in some cases [34, 35].

Since $p - 1$ *zeroes* are always required between the *ones* I call them *bound zeroes* which seem to represent the mutual statistical exclusions. In the case of the free Fermions carrying the flux 2π in units where $e = \hbar = c = 1$ so that the flux quanta $\Phi_0 = 2\pi$, the minimum separation of the momenta is $2\pi/L$ which can be considered as the mutual Pauli exclusion. In this case the minimum separation is $2\pi\lambda/L$; therefore, it is natural to assign $2\pi\lambda$ flux attached to the pseudo-particles. The remaining *zeroes* are called *unbound zeroes*. The q consecutive unbound *zeroes* in the condensate of the pseudo-particles constitute a single hole excitation. Thus, if a pseudo-particle is removed from the ground state condensate, then there are p *unbound zeroes* in place where the *one* is removed. This state is forbidden if $q \neq 1$. In general a minimum of q *ones* must be removed so that they leave behind at least pq *unbound zeroes* which break up into p holes. From the view point of the particles (holes) the change in the number of available single particle states is p $(-q)$ while the change in the number of quasiparticles (quasiholes) in the system is $-q$ (p). Therefore, the statistical parameter g for the quasiparticle (quasihole) is $g = p/q = \lambda$ ($g = q/p = 1/\lambda$). This inverse relationship between the statistical parameters of the particles and holes is essentially the Chern-Simons duality. To summarize the fractional exclusion statistics, q *particle excitations are accompanied by p hole excitations*.

The configurations constructed above are representations of the diagrams of partitions $\mathcal{D}(\kappa)$ introduced in Sec. 7.3. The part κ_j corresponds to the displacement of jth quantum number from the ground state (i.e., $I_j - I_j^0$ if $I_1 > I_2 > \ldots > I_N$). The excitations given by κ include only the states with non-negative displacements (i.e., k_j moved only to the right) and all the other states are obtained by global Galilean transformations. Therefore, each row (column) in the diagram corresponds to the

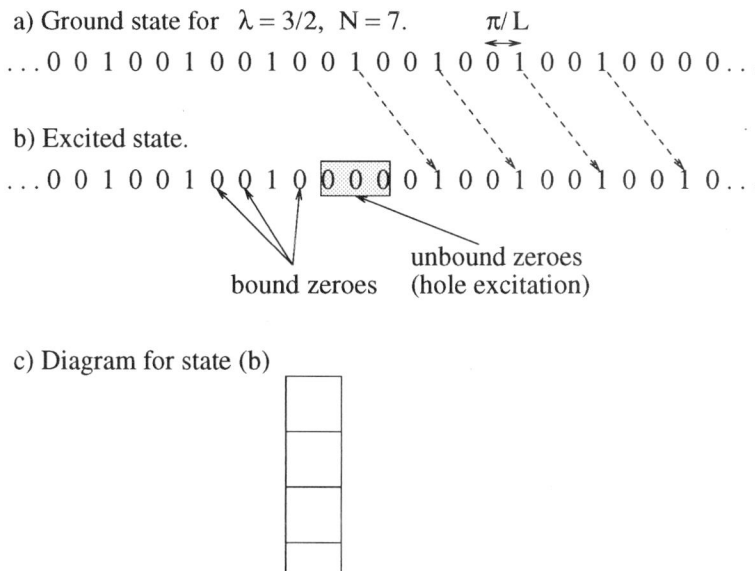

a) Ground state for $\lambda = 3/2$, N = 7. π/L

...0 0 1 0 0 1 0 0 1 0 0 1 0 0 1 0 0 1 0 0 1 0 0 0 0...

b) Excited state.

...0 0 1 0 0 1 0 0 1 0 0 0 0 0 0 1 0 0 1 0 0 1 0 0 1 0...

                                        unbound zeroes
                bound zeroes   (hole excitation)

c) Diagram for state (b)

Fig. 7.5. Illustration of ground state and excited state motif structure for $\lambda = 3/2$ and $N = 7$.

particle (hole) excitations of the CSM. Refer to Fig. 7.5 for an illustration.

Following Yang and Yang [110] and Sutherland [95] I can also construct the thermodynamics [34, 7]. In the thermodynamic limit it is convenient to define the hole distribution $\rho_h(k)$ and the particle distribution function $\rho_p(k)$. The function $\rho_p(k)$ are given by the solutions of Eq. (7.14) while $\rho_h(k)$ by the corresponding complementary equations given by the unused quantum numbers I_j; thus, they satisfy $1 = \rho_h(k) + \lambda \rho_p(k)$. This equation states that one hole and $1/\lambda$ particles (or λ holes and one particle) have equal weight in occupying the volume in k-space. Hence, for $\lambda \neq 1$ the particle-hole symmetry is broken; and in the bosonic case ($\lambda = 0$) the symmetry is maximally broken. With a proper normalization one can equally state that $1 = (1/\lambda)\rho_h(k) + \rho_p(k)$. This particle-hole duality is the essence of the Chern-Simons duality.

The thermodynamic function is given by $\Omega = E - TS - \mu N$, where the energy (E), the particle number (N), and the entropy (S) are given by

$$E = V \int dk \rho_p(k) e(k), \qquad (7.38)$$

$$N = V \int dk \rho_p(k), \qquad (7.39)$$

$$S = V \int dk \{ (\rho_h + \rho_p) \log(\rho_h + \rho_p) - \rho_h \log \rho_h - \rho_p \log \rho_p \}. \qquad (7.40)$$

Here, V is the volume of the system. By minimizing $\Omega(\rho_p, \rho_h)$ with respect to the density functions the following relation can easily be obtained

$$(1 - \lambda \rho_p(k))^\lambda (1 + (1 - \lambda)\rho_p(k))^{1-\lambda} = \rho_p(k) e^{(\epsilon(k) - \mu)/T}. \qquad (7.41)$$

This equation is also obtained by Wu using a different method [107].

7.5 Dynamical density-density correlation function

In this section I show that the dynamical density-density correlation function (DDDCF) can be calculated exactly for the CSM using the known properties of Jack polynomials. The DDDCF is defined by

$$\begin{aligned}
{}_N\langle 0 | \rho(x', t') \rho(x, t) | 0 \rangle_N &= {}_N\langle 0 | e^{iH_N t'} \rho(x') e^{-iH_N t'} e^{iH_N t} \rho(x) e^{-iH_N t} | 0 \rangle_N \\
&= {}_N\langle 0 | \rho(x') e^{-i(H_N - E_N^0)(t' - t)} \rho(x) | 0 \rangle_N, \qquad (7.42)
\end{aligned}$$

where the *reduced* density operator $\rho(x) = (1/L) \sum_j \delta(x - x_j) - N/L$ and $|0\rangle_N$ the normalized N-particle ground state.

The first step in calculating the DDDCF is to expand $\rho(x)|0\rangle_N$ in terms of the eigenstates of the CSM (i.e., Jack polynomials). The delta function $\delta(x)$ is a periodic function with a period L and thus can be expressed as a Fourier sum $(1/L) \sum_{m=-\infty}^{+\infty} \exp(i2\pi xm/L)$. Therefore, I can write $\rho(x)$ as follow

$$\rho(x) = \frac{1}{L} \sum_{m=1}^{\infty} (z^m p_{-m} + z^{-m} p_m), \qquad (7.43)$$

where $z = \exp(i2\pi x/L)$, $z_j = \exp(i2\pi x_j/L)$, and $p_m = \sum_{j=1}^N z_j^m$. The power sum p_m can be expanded in terms of Jack polynomials using the identity Eq. (7.23).

Using the orthogonality relation Eq. (7.21) and its extension Eq. (7.22), I obtain the following expression for the DDDCF

$${}_N\langle 0 | \rho(x, t) \rho(0, 0) | 0 \rangle_N = \frac{1}{L^2} \frac{2}{\lambda^2} \sum_\kappa \frac{|\kappa|^2}{j_\kappa^\lambda} \frac{([0']_\kappa^\lambda)^2 [N]_\kappa^\lambda}{[N + 1/\lambda - 1]_\kappa^\lambda} \cos(2\pi |\kappa| x/L) e^{-itE_\kappa}, \quad (7.44)$$

where $E_\kappa = (2\pi/L)^2 \sum_{j=1}^N (\kappa_j^2 + \lambda(N + 1 - 2j)\kappa_j)$. The coefficient $[0']_\kappa^\lambda$ in Eq. (7.44) vanishes unless the diagram $\mathcal{D}(\kappa)$ has no more than p columns of length longer than q and q rows of length longer than p. In other words, the intermediate states contributing to the DDDCF has precisely p hole and q particle excitations. This is a conclusive evidence of the ideal fractional exclusion statistics the CSM quasiparticles and quasiholes obey.

The DDDCF in the thermodynamic limit greatly simplifies as is shown in Appendix A; and it is given by [34, 35]

$$\langle 0|\rho(x,t)\rho(0,0)|0\rangle = C \prod_{i=1}^{q} \left(\int_0^\infty dx_i \right) \prod_{j=1}^{p} \left(\int_0^1 dy_j \right) Q^2 F(q,p,\lambda|\{x_i,y_j\}) \cos(Qx) e^{-iEt},$$

(7.45)

where Q and E, the total momentum and energy, are given in units of $\hbar$ and $\hbar^2/2m$ by

$$Q = 2\pi\rho_0 \left(\sum_{j=1}^{q} x_j + \sum_{j=1}^{p} y_j \right),$$

(7.46)

$$E = (2\pi\rho_0)^2 \left(\sum_{j=1}^{q} \epsilon_P(x_j) + \sum_{j=1}^{p} \epsilon_H(y_j) \right),$$

(7.47)

with $\rho_0 = N/L$, $\epsilon_P(x) = x(x+\lambda)$ and $\epsilon_H(y) = \lambda y(1-y)$. $x_j(\epsilon_P)$ and $y_j(\epsilon_H)$ are normalized momentum(energy) of the quasiparticles and the quasiholes, respectively. The normalization constant C is given by

$$A(m,n,\lambda) = \frac{\Gamma^m(\lambda)\Gamma^n(1/\lambda)}{\prod_{i=1}^{m} \Gamma^2(p - \lambda(i-1)) \prod_{j=1}^{n} \Gamma^2(q - (j-1)/\lambda)}$$

$$\times \prod_{j=1}^{n} \left(\frac{\Gamma(q - (j-1)/\lambda)}{\Gamma(1 - (j-1)/\lambda)} \right)^2,$$

(7.48)

$$C = \frac{\lambda^{2p(q-1)}\Gamma^2(p)}{2\pi^2 p! q!} A(q,p,\lambda).$$

(7.49)

Finally, the form factor $F(q,p,\lambda|\{x_i,y_j\})$ is given by

$$F(m,n,\lambda|\{x_i,y_j\}) = \prod_{i=1}^{m} \prod_{j=1}^{n} (x_i + \lambda y_j)^{-2} \frac{\left(\prod_{i<j}(x_i - x_j)^2 \right)^\lambda \left(\prod_{i<j}(y_i - y_j)^2 \right)^{1/\lambda}}{\prod_{i=1}^{m} \epsilon_P(x_i)^{1-\lambda} \prod_{j=1}^{n} \epsilon_H(y_j)^{1-1/\lambda}}.$$

(7.50)

The form factor has been conjectured by Haldane based on the clues given by the works of Simons et. al., Galilean invariance, and $U(1)$ conformal field theory [49]. A less general form of the DDDCF at integer values of λ has also been reported [75].

The region of support in the energy-momentum space for the DDDCF at $\lambda = 5/3$ corresponds to the shaded area in Fig. 7.6. It is obtained by convoluting the dispersion relations of the q quasiparticles and p quasiholes as given by Eqs. (7.46) and (7.47). At low-energies $2p$ distinct sectors indicated by darker shade emerge as expected of 1D metallic system. (A generic 1D system, however, will have an infinite number of these low-energy sectors.)

It is also worth noting that there is a qualitative difference between the finite and the infinite system. The intermediate states represented by the diagrams with

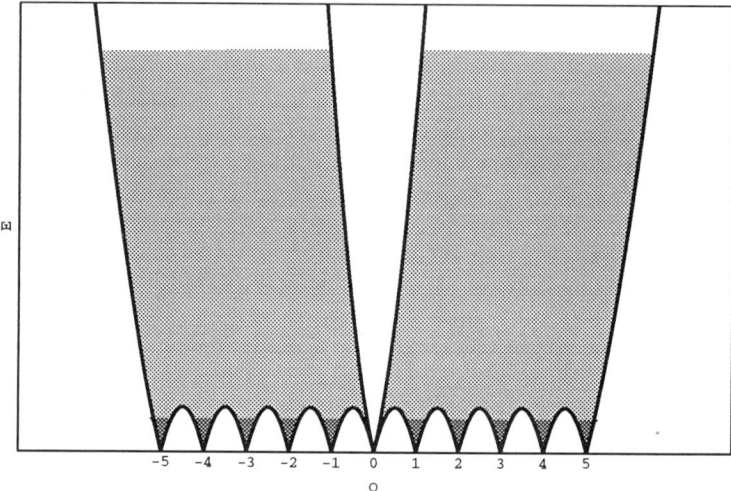

Fig. 7.6. The shaded area is the region of support for the dynamical density-density correlation function at $\lambda = p/q = 5/3$ in the energy-momentum space. The energy E is in some arbitrary unit and the momentum Q in units of $2\pi\rho_0$. The $2p$ distinct regions with darker shade corresponds to the low-energy sectors of the model.

$|\kappa_j - \kappa_k|$ and $|\kappa'_j - \kappa'_k|$ of order $\mathcal{O}(N)$ are the only states that contribute to the DDDCF in the thermodynamic limit (i.e., N, $L \to \infty$ with N/L fixed). I call this a super-selection rule [34]. (There can be violations of this rule at the low-energy limit as discussed in Appendix A.) Perhaps, the phase transition that occurs in 2D QCD as $N \to \infty$ [28] is related to the CSM super-selection rule.

7.6 One-Particle Green's Function

Unlike the DDDCF calculated in the previous section, evaluation of the one particle Green's function depends on the actual statistics of the real particles since it involves the reordering of the particles. For example, the one-particle density matrix (the static limit of the Green's function) for the $\lambda = 1$ case is trivially simple if the ground state Ψ_0 is taken to be the fermionic wave function. On the other hand, if the wavefunction is taken to describe a bosonic system, the calculation gets quite complicated and different from its fermionic counterpart as shown in [74].

As described in section 7.4 I take the modified CSM given by Eq. (7.32) as one-dimensional anyon system with the exchange phase given by $\exp(i\lambda\pi)$ and Eq. (7.33) as the corresponding wavefunctions. If the mth particle is removed from the ground state of $N + 1$ particles, the remaining term after factoring out the ground state wavefunction of the N particle system is

$$\left(\prod_{j(<m)} (z_j - z_m)^\lambda\right) \left(\prod_{j(>m)} (z_m - z_j)^\lambda\right) \left(\prod_{k \neq m} z_k^{-\lambda/2}\right) z_m^{-\lambda N/2} e^{i(m-1)\pi\lambda}, \qquad (7.51)$$

where the last term is determined from the property of the ordering function φ given in Eq. (7.34). Therefore, up to an overall phase factor that does not depend on the position of the removed particle, the destruction operation $\Psi(x)$ on the ground state of $N + 1$ particles is given by

$$\Psi(x)|0\rangle_{N+1} = \frac{A_{N+1}}{A_N} z^{-\lambda N/2} \prod_{j=1}^{N} (z - z_j)^\lambda z_j^{-\lambda/2}|0\rangle_N, \qquad (7.52)$$

where $z = \exp(i2\pi x/L)$ and $A_N^2 = (1/L)^N \Gamma^N(1+\lambda)/\Gamma(1+\lambda N)$. The statistical phase arising from the ordering function makes the destruction operation symmetric with respect to the permutations of $\{z_j\}$. The symmetric function $\prod_{j=1}^{N}(z - z_j)^\lambda$ can then be expanded in term of Jack polynomials using Eqs. (7.24) and (7.25).

The hole propagator part of the one-particle Green's function is defined as

$$\begin{aligned}
_{N+1}\langle 0|\Psi^\dagger(x', t')\Psi(x, t)|0\rangle_{N+1} &= {}_{N+1}\langle 0|e^{iH_{N+1}t'}\Psi^\dagger(x')e^{-iH_N t'} e^{iH_N t}\Psi(x)e^{-iH_{N+1}t}|0\rangle_{N+1} \\
&= {}_{N+1}\langle 0|\Psi^\dagger(x')e^{-i((H_N - E_N)-\mu)(t'-t)}\Psi(x)|0\rangle_{N+1}, \quad (7.53)
\end{aligned}$$

where the chemical potential $\mu = E_{N+1} - E_N$. By expanding $\Psi(x)$ in terms of Jack polynomials and using the orthogonality relation (7.21) the propagator at finite N

and L is evaluated to be

$$
\begin{aligned}
_{N+1}\langle 0|\Psi^\dagger(x,t)\Psi(0,0)|0\rangle_{N+1} &= (N+1)\left(\frac{A_{N+1}}{A_N}\right)^2 \sum_{\boldsymbol{\kappa}} \frac{\lambda^{-2|\boldsymbol{\kappa}|}}{j_{\boldsymbol{\kappa}}^\lambda} \frac{(\{-\lambda\}_{\boldsymbol{\kappa}}^\lambda)^2 [N]_{\boldsymbol{\kappa}}^\lambda}{[N+1/\lambda-1]_{\boldsymbol{\kappa}}^\lambda} \\
&\quad \times \ e^{i2\pi(|\boldsymbol{\kappa}|-\lambda N/2)x/L} e^{-i(E_{\boldsymbol{\kappa}}-\mu)t},
\end{aligned}
\tag{7.54}
$$

where the additional $(N+1)$ factor comes from the freedom of choosing one of $N+1$ available particles to destroy and create. The coefficient $\{-\lambda\}_{\boldsymbol{\kappa}}^\lambda$ vanishes unless the diagram $\mathcal{D}(\boldsymbol{\kappa})$ has at most $q-1$ rows of length greater than p and p columns of length greater than $q-1$. Therefore, the intermediate states for the propagator is spanned by $q-1$ quasiparticles and p quasiholes. This is a very important result. *The exclusion statistics of the quasiparticles and quasiholes is completely consistent with the anyon statistics of the real particles.* [34, 35]

Taking the thermodynamic limit of Eq. (7.54) is almost identical to that of Eq. (7.44) and is given by [34, 35]

$$
\begin{aligned}
\langle 0|\Psi^\dagger(x,t)\Psi(0,0)|0\rangle &= \rho_0 D \prod_{i=1}^{q-1}\left(\int_0^\infty dx_i\right)\prod_{j=1}^{p}\left(\int_0^1 dy_j\right) F(q-1,p,\lambda|\{x_i,y_j\}) \\
&\quad \times \ e^{i((Q-Q_0)x-(E-\mu)t)},
\end{aligned}
\tag{7.55}
$$

where the chemical potential $\mu = (\pi\lambda\rho_0)^2$ and the back flow $Q_0 = \pi\lambda\rho_0$. $F(q-1,p,\lambda|\{x_i,y_j\})$ is still given by Eq. (7.50) and D by

$$
D = \frac{\lambda^{2p(q-1)}\Gamma^2(p)}{\Gamma(\lambda)(q-1)!p!} A(q-1,p,\lambda).
\tag{7.56}
$$

Q and E are same as before except for the number of x_j's. At integer values of λ (i.e., $q=1$ case where only quasiholes are excited), based on the equal-time results of Forrester [21] Haldane made a conjecture [46] which agrees with this formula. The regions of support for the hole propagator is given by the shaded area in $Q > 0$ (or $Q < 0$) in Fig. 7.6. There are also shifts in E and Q by $-\mu$ and $-Q_0$, respectively.

It is also interesting to consider the following function

$$
\Psi_m^\lambda(x) = \prod_{i<j}(z-z_j)^m,
\tag{7.57}
$$

where m is a positive integer. This function can be expanded in terms of Jack polynomials using Eq. (7.24) with coefficients containing the term $\{-m\}_{\boldsymbol{\kappa}}^\lambda$ which vanishes unless the diagram $\mathcal{D}(\boldsymbol{\kappa})$ has no more than m columns and no rows longer than m. Therefore, $\Psi_m^\lambda(x)$ acting on the ground state creates exactly m quasiholes. $\Psi_m^\lambda(x)$ is a generalization of the one-hole state given in Eq. (7.9). The propagator can easily be calculated and its form factor is given by $F(0,m,\lambda|\{y_j\})$. This result is consistent with the conjecture [34, 35] that the minimal form factor for any two point correlation functions whose intermediate states contain only n quasiparticle and m quasiholes is given by $F(n,m,\lambda|\{x_i,y_j\})$.

7.7 Low-energy effective theory and Coupled Fractional Quantum Hall Edge States

Following the standard method in Luttinger liquid theory [37, 38] I construct an effective low-energy model for the 1D anyon system. When λ is an integer this system is equivalent to a coupled system of left- and right-moving edge states of the FQHE. The excitations on a single edge of the FQH fluid moves only in one direction because of the externally applied magnetic field; and they have been rather thoroughly studied using the so called chiral Luttinger liquid theory [104] which is intrinsically anomalous [113]. When there is an extra edge on the FQH droplet (e.g. a strip, annulus, or cylindrical geometry instead of the disk geometry) and when the edges are close enough, new phase space opens up as a result of the fractional charge transfer between the edges. There actually have been many suggestions that the CSM is related to the edge states [48]. In this section I show in another way that the CSM is an exactly solvable 1D anyon system by calculating the correlation functions of the effective model and finding exact agreement with those of the CSM in the long-wavelength limit.

A general gauge invariance argument [101] can be used to map out the qualitative structure of the excitation spectra in the energy-momentum space. I use the cylindrical geometry [101] with coordinates x and y defined as shown in Fig. 7.7. I put some incompressible fluid perhaps made up of λ anyons on the surface of the cylinder with a confining potential in y-direction such that two edges are created along x-direction. (Of course, when λ is not an integer the edges are not related to the FQHE edges which are more complicated composite type [104].) The cylinder with circumference L is also pierced by a thin solenoid along its longitudinal axis, thus inducing a magnetic flux $\Phi = AL$ through its cross-section and a vector potential $\vec{A} = A\hat{x}$ on its surface. There is also magnetic fields $\vec{B}$ normal to the entire surface of the cylinder. Since the left- and right-moving sectors of the CSM completely decouple in the low-energy limit, this geometrical construction is in fact equivalent to the CSM and is useful to show intuitively how the low-energy sectors are related.

The fictitious flux applied to the cylinder is used as a passive device for mapping out the regions of low-energy excitation sectors of the coupled anyon edges. By virtue of the gauge invariance the energy is degenerate at flux equal to $2\pi n$ (in units where $\hbar = c = 1$) which corresponds to the momentum of $2\pi n\rho_0$. The adiabatical change in flux from 0 to 2π induces an elementary excitation carrying charge $-q/p$ and flux -2π (i.e., a quasihole) to move from one edge to the other; and, in between, it costs finite energy to the system since the bulk is incompressible. In fact the real space configuration of an incompressible droplet is equivalent to a "Fermi sea." [23] The edges of the fluid coincide with the locus of points in which Fermi energy crosses the external confining potential. The only degrees of freedom left in "Fermi sea" in the low-energy limit is the "Fermi surface" fluctuation. This "Fermi sea" obeys the fractional exclusion statistics. Since there is a symmetry along x-direction, the fluid can be regarded as an 1D "Fermi sea" and the left- and the right-edges as two "Fermi

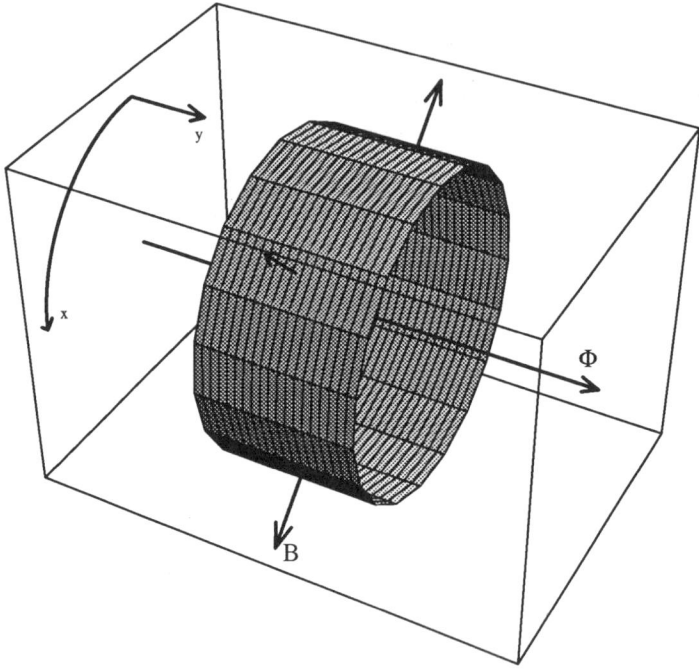

Fig. 7.7. Cylindrical geometry for the anyon system. Two oppositely-moving edges along x-axis are separated by an incompressible anyon fluid. The cylinder is pierced by a solenoid which induces a magnetic flux Φ through its cross-section.

points."

In light of the observations made so far it is reasonable to model the coupled edges by a full (non-chiral) 1D anyon system. First, I construct the one-dimensional anyon creation operator as follow

$$\Psi_\lambda^\dagger(x) = \Psi_B^\dagger(x)e^{i\lambda\theta(x)}, \tag{7.58}$$

where $\Psi_B^\dagger$ is the boson creation operator and λ the statistical parameter. The operator $e^{i\lambda\theta(x)}$ is the so called "disorder operator" [58] that creates a kink (or vortex) of size $\pi\lambda$ at position x and is defined as

$$\theta(x) = \pi \int_{-\infty}^{x} \rho(x')dx', \tag{7.59}$$

where $\rho(x)$ is the density operator. Therefore, $\Psi_\lambda^\dagger(x)$ is a composite operator that creates a boson plus a vortex; so, it is an anyon creation operator.

The boson creation operator $\Psi_B^\dagger(x)$ for a system with multiple low-energy sectors has previously been constructed [38] and is given by

$$\Psi_B^\dagger(x) \approx \rho_0^{1/2} \sum_{m=-\infty}^{+\infty} e^{i2m\theta(x)}e^{i\phi(x)}, \tag{7.60}$$

where the phase field $\phi(x)$ is defined by the following canonical commutation relation: $[\phi(x), \rho(x')] = i\delta(x - x')$. The multi-sector density operator is also given by

$$\tilde{\rho}(x) = \rho(x) \sum_{m=-\infty}^{+\infty} e^{i2m\theta(x)}. \tag{7.61}$$

Here, the sectors are connected by the operator $\exp(i2m\theta(x))$ which creates a vortex of size $2\pi m$. This dynamical device now replaces the passive device I previously used to map the system from one sector to the next by supplying external magnetic flux to the cylinder.

The fields $\theta(x)$, $\phi(x)$ and $\rho(x)$ can now be expressed in terms of Tomonaga boson operators $b^\dagger$ and b as [38, 83]

$$\theta(x) = \theta_0 + \pi\rho_0 x + ie^\varphi \sum_{k\neq 0} \frac{1}{k}\left|\frac{\pi k}{2L}\right|^{1/2} e^{-ikx}(b_{-k}^\dagger + b_k) \tag{7.62}$$

$$\phi(x) = \phi_0 - ie^{-\varphi} \sum_{k\neq 0}\left|\frac{\pi}{2kL}\right|^{1/2} e^{-ikx}(b_{-k}^\dagger - b_k) \tag{7.63}$$

$$\rho(x) = \rho_0 + e^\varphi \sum_{k\neq 0}\left|\frac{k}{2\pi L}\right|^{1/2} e^{-ikx}(b_{-k}^\dagger + b_k). \tag{7.64}$$

I assume here that the most important interaction is the statistical interaction; so, I can set the Bogoliubov parameter $e^{2\varphi} = 1/\lambda$.

Now, $\Psi_\lambda^\dagger(x)$ can easily be shown to satisfy the following anyon commutation relation

$$\Psi_\lambda^\dagger(x)\Psi_\lambda^\dagger(x') = e^{i\pi\lambda\mathrm{sgn}(x'-x)}\Psi_\lambda^\dagger(x')\Psi_\lambda^\dagger(x) \quad \text{for } x \neq x'. \tag{7.65}$$

Eq. (7.65) can thus be used as a defining relation for the one-dimensional anyons.

The Hamiltonian is diagonal in Tomonaga boson operators and is given as usual by

$$H = \hbar v_s \sum_{q \neq 0} |q| b_q^\dagger b_q, \tag{7.66}$$

where v_s is the sound velocity.

Using the operator identities

- (i) $e^A e^B = \exp(e^\alpha - 1)e^B e^A$ if $[A, B] = \alpha B$,

- (ii) $e^A e^B = \exp(-[A, B])e^B e^A$ and $e^{A+B} = \exp(-[A, B]/2)e^A e^B$
  if $[[A, B], A] = [[A, B], B] = 0$,

the following correlation functions are calculated

$$\langle \hat{\rho}(x,t)\hat{\rho}(0,0) \rangle \approx \rho_0^2 \left(1 - \frac{\lambda^{-1}}{(2\pi\rho_0)^2} \left(\frac{1}{(\xi^-)^2} + \frac{1}{(\xi^+)^2} \right) \right.$$
$$\left. + \sum_{m=1}^{\infty} A_m \left(\frac{1}{\xi^+\xi^-} \right)^{m^2/\lambda} \cos(2\pi\rho_0 m x) \right), \tag{7.67}$$

$$\langle \Psi_\lambda^\dagger(x,t)\Psi_\lambda(0,0) \rangle \approx \rho_0 \sum_{m=-\infty}^{\infty} B_m \left(\frac{1}{\xi^+} \right)^{(m+\lambda)^2/\lambda} \left(\frac{1}{\xi^-} \right)^{m^2/\lambda} e^{i(2\pi\rho_0(m+\lambda/2)x+\mu t)}, \tag{7.68}$$

where $\xi^\pm = x \mp v_s t$, μ the chemical potential and the coefficients A_m (B_m) are regularization-dependent constants.

The Green's function $\langle \Psi_\lambda^\dagger(x,t)\Psi_\lambda(0,0) \rangle$ for the sector $m = 0$, where the charge transfer from one edge to the other is forbidden, is given by only the right-movers (or only the left-movers if $\Psi_\lambda^\dagger$ were properly redefined) even though the anyon creation operator $\Psi_\lambda^\dagger(x)$ contains both the right- and the left-moving bosonic modes (i.e., $b_k^\dagger$ and b_k for both $k > 0$ and $k < 0$). This chiral sector emerges naturally in this theory without explicitly imposing the chirality condition.

In an *isolated* chiral theory coupled to a gauge field the charge is in general not conserved and because of this the theory is known to be anomalous and not physical. As far as the isolated chiral theory is concerned the only physical sector is the chiral sector where no charge transfer between the edges is possible. One can, however, consider an "almost" chiral theory which describes a system of two chiral edge states that are independent except for the charge leakage from one side to the other and vice versa. For the sake of definiteness I concentrate on the right edge and propose the "almost" chiral system with the following Hamiltonian and the field operators,

$$H^R = \hbar v_s \sum_{q>0} q b_q^\dagger b_q, \tag{7.69}$$

$$\theta^R(x) = \theta_0 + \pi\rho_0^R x - \frac{i}{\sqrt{\lambda}}\sum_{k>0}\frac{1}{k}\left|\frac{\pi k}{2L}\right|^{1/2}\left(e^{ikx}b_k^\dagger - e^{-ikx}b_k\right), \qquad (7.70)$$

$$\phi^R(x) = \phi_0 - i\sqrt{\lambda}\sum_{k>0}\left|\frac{\pi}{2kL}\right|^{1/2}\left(e^{ikx}b_k^\dagger - e^{-ikx}b_k\right), \qquad (7.71)$$

$$\Psi_\lambda^{R\dagger}(x) \approx \sqrt{\rho_0^R}\sum_m e^{i2(m+\lambda/2)\theta^R(x)}e^{i\phi^R(x)}. \qquad (7.72)$$

The chiral system is constructed only with the right-moving Tomonaga bosons. The Green's function in this case is given by

$$\langle\Psi_\lambda^{R\dagger}(x,t)\Psi_\lambda^R(0,0)\rangle \approx \rho_0^R\sum_{m=-\infty}^{\infty}C_m\left(\frac{1}{x-v_s t}\right)^{(m+\lambda)/\lambda}e^{i(2\pi\rho_0^R(m+\lambda/2)x+\mu^R t)}. \qquad (7.73)$$

The $m = 0$ sector here is equivalent to the corresponding chiral sector of the non-chiral model. As expected only the right-movers contribute to the Green's function for all the other sectors.

Asymptotic expansions of the correlation functions of the CSM have been calculated in Appendix B; and they agree with Eq. (7.67) and Eq. (7.68). This agreement between the correlation functions of the explicitly constructed anyon model and the CSM shows that the first- and the second-quantized construction of the anyons are in fact equivalent in the long-wavelength limit.

7.8   Finite-size Scaling and correlation exponents

There is another extremely elegant and powerful way of obtaining the exponents of the correlation functions in the long-wavelength limit. If the dispersion relations of the elementary excitations of the one-dimensional quantum model in the low-energy limit is linear (i.e., has Lorentz as well as Galilean invariance so that the space and time variables are on equal footing), the principle of conformal invariance is applicable and, in particular, the finite-size corrections to the energy and momentum become universal and are directly related to the exponents of the correlation functions [12].

The finite-size scaling has previously been applied to the CSM by Kawakami and Yang [66] without the benefit of recently uncovered knowledge [46, 34], namely that due to the ideal fractional statistics the CSM particles obey the intermediate states for the DDDCF and the Green's function are spanned by finite number of the elementary excitations. The shaded regions in Fig. 7.6 are the relevant sectors for the correlation functions (for the Green's function only the positive or negative momentum sectors contribute). A complete set of relevant exponents is obtained using the finite-size scaling in this section.

Let $(p_L, h_L|p_R, h_R)$ be a label for a low-energy sector spanned by p_L (h_L) left-moving and p_R (h_R) right-moving quasiparticles (quasiholes). The relevant sectors for the DDDCF are $(0, n|q, p - n)$ and $(q, p - n|0, n)$ and for the Green's function $(0, n|q, p - n)$ or $(q, p - n|0, n)$ where $n = 0, 1, \ldots, p$. First, for the DDDCF the lowest-energy state in the sector m is characterized by m quasiholes on right (or left) side of the "Fermi sea" transfered to the left (or right). Hence, the pseudo-momenta k_j for this sector is given in terms of the ground-state pseudo-momenta k_j^0 as $k_j = k_j^0 + 2\pi m/L$; and the energy and momentum by

$$E_m = E_0 + \frac{2\pi v_s}{L}\frac{m^2}{\lambda}, \tag{7.74}$$

$$P_m = 2\pi m \rho_0, \tag{7.75}$$

where the sound velocity $v_s = 2\pi\lambda\rho_0$. The scaling dimension $x = h^+ + h^-$ of the density operator and its conformal spin $s = h^+ - h^-$ are therefore given by $x = m^2/\lambda$ and $s = 0$, where $h^\pm$ are right (left) conformal weights and $2h^\pm$ are the exponents that actually appear in the correlation functions for the right(left) movers. Thus, $2h^+ = 2h^- = m^2/\lambda$ in agreement with the exponents found in the previous section and in Appendix B.

Now, for the hole propagator the energy and momentum of the sector m characterized by m quasihole transfers from the left to right "Fermi points" and thus the pseudo-momenta $k_j = k_j^0 + 2\pi m/L$, where $j = 1, \ldots, N-1$, are

$$E_m = E_0 - \mu + \frac{2\pi v_s}{L}\left(\frac{m^2}{\lambda} - n + \frac{\lambda}{2}\right), \tag{7.76}$$

$$P_m = 2\pi(n - \lambda/2)\rho_0 + \frac{2\pi}{L}\left(\frac{\lambda}{2} - n\right). \tag{7.77}$$

The chemical potential $\mu = (\pi\lambda\rho_0)^2$ is associated with the particle destruction. I remove the rightmost pseudo-particle with $k_N^0 = \pi\lambda(N-1)/L$ from the ground state condensate and do not introduce any separate selection rules in contrast to Ref. [66]. In this case the right and left conformal weights are different (due to the non-zero conformal spin $s = \lambda/2 - n$) and are given by $2h^+ = (m - \lambda)^2/\lambda$ and $2h^- = m^2/\lambda$ as expected. The right and left conformal weights would be switched if the excitations were caused by the removal of the leftmost pseudo-particle at $-\pi\lambda(N-1)/L$.

Apparently, the long-range interaction does not destroy the conformal invariance of the CSM in the long-wavelength limit. This is expected from the linear dispersion relations of the CSM quasiparticles and quasiholes in this limit.

7.9 Lattice Models

The Haldane-Shastry model (HSM) corresponds to a lattice generalization of the CSM at $\lambda = 2$ and has the following Hamiltonian

$$H = J_0 \sum_{i<j} \frac{\vec{S}_i \cdot \vec{S}_j}{d^2(i-j)}, \tag{7.78}$$

where $d(m) = (N/\pi)|\sin(\pi m/N)|$ and $\vec{S}_j$ is the $SU(2)$ spin operator acting on site j. The model possesses a quantum symmetry called Yangian [42] and exhibits supermultiplets structures whose spin contents are exactly reproducible from an asymptotic limit of the thermodynamic Bethe-ansatz equations [32]. Furthermore, there is one-to-one correspondence between the highest weight states of the Yangian supermultiplets and the states of the CSM (i.e., they satisfy the same eigenvalue equations.) A further generalization to $SU(n)$ case has also been accomplished in [31, 67]. The lattice CSM model at even (odd) integer values of λ can be mapped to bosonic (fermionic) spinless t-J model [31].

The method for evaluating the correlation functions of the CSM presented in this chapter is directly applicable to its lattice cousins in some limited cases. The Galilean invariance is broken in the lattice models and is replaced with much weaker lattice translation symmetry which induces appearance of the Brillouin zone boundaries. If the elementary excitations created by a local operator acting on the ground state do not cross the Brillouin zone boundaries and the excited states are the Yangian highest weight states, then the corresponding correlation functions of HSM are identical to that of the CSM.

7.10 Conclusion

In this chapter the fractional exchange and exclusion statistics are studied using the exactly solvable Calogero-Sutherland model; and they are found to be mutually consistent. I have shown that the interaction giving rise to the fractional exclusion statistics for the elementary excitations of a given condensed matter system can in fact be treated as the statistical gauge field carried by the real particles making up the system. This was done by calculating the exact dynamical density-density correlation function and one-particle Green's function using Jack symmetric polynomials and examining the intermediate states contributing to the correlation functions. I find that the intermediate states for $\lambda = p/q$ CSM are spanned by q quasiparticles and p quasiholes for the density-density correlation function and $q - 1$ quasiparticles and p quasiholes for the hole propagator and thereby show that the quasiparticles indeed carry charge 1 and flux $2\pi\lambda$ and the quasiholes charge $-1/\lambda$ and flux -2π as first suggested by Haldane [16].

I also construct explicit multi-sector anyon operators in analogy with Haldane's harmonic-fluid [38] and calculate their correlation functions which agree with those of the CSM which corresponds to the first-quantized construction of anyons. Therefore, the CSM at odd-integer coupling constant describes the edge states of the fractional quantum Hall droplet corresponding to the Laughlin states as suggested in Ref. [48].

There are some interesting open problems:

- How can one rigorously construct 1D fractional exchange statistics (in general non-Abelian) using the Yang-Baxter equation ?

- For the CSM with the coupling constant other than $\lambda = 1/2$, 1, and 2, are there any corresponding random matrices?

- How can one generalize Jack polynomials to the $SU(n)$ case?

Presumably in the systems with internal degrees of freedom the string structure will play a role but what role is unclear. I hope to see some of these questions answered in the future.

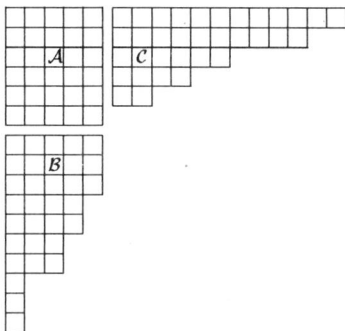

Fig. 7.8. A diagram $\mathcal{D}$ that contributes to the DDDCF at $\lambda = p/q = 5/6$ is divided into three subdiagrams $\mathcal{A}$, $\mathcal{B}$, and $\mathcal{C}$ for convenience.

7.A How to take the thermodynamic limit

The thermodynamic limit of the DDDCF given by Eq. (7.44) greatly simplifies by the fact that the coefficient $[0']^\lambda_\kappa$ vanishes unless the diagram $\mathcal{D}(\kappa)$ has at most p rows of length longer than q and q columns of length longer than p. As shown in Section 7.4 the minimal charge neutral excitations in the CSM consist of q quasiparticles and p quasiholes; therefore, the intermediate states that contribute to the correlation function have exactly q quasiparticle and p quasihole excitations.[2]

The contributing diagrams $\mathcal{D}(\kappa)$, for convenience, are divided into three subdiagrams $\mathcal{A}(\kappa) = \{(i,j),\ 1 \le i \le p,\ 1 \le j \le q\}$, $\mathcal{B}(\kappa) = \{(i,j),\ 1 \le j \le q,\ q+1 \le i \le \kappa'_j\}$ and $\mathcal{C}(\kappa) = \{(i,j),\ 1 \le i \le p,\ p+1 \le j \le \kappa_i\}$ as illustrated in Fig. 7.8. The factors expressed in terms of the generalized factorial over $\mathcal{D}(\kappa)$ are evaluated for each of the subdiagrams, separately. Then, later the each subfactors are multiplied to obtain to full factors over the diagram $\mathcal{D}$.

First, consider the following factor that appears in the DDDCF

$$F_1 = \prod_{(i,j) \in \kappa} \left(\frac{N + (j-1)/\lambda - (i-1)}{N + j/\lambda - i} \right). \qquad (7.A.79)$$

---

[2]The non-vanishing diagrams with less than p rows and q columns are interpreted as having some of the quasiparticles and quasiholes in their unexcited modes.

F_1 over $\mathcal{A}$ simplifies in the thermodynamic limit to

$$F_1(\mathcal{A}) = \prod_{j=1}^{p} \prod_{i=1}^{q} \left(\frac{N + (j-1)/\lambda - (i-1)}{N + j/\lambda - i} \right) \stackrel{N \to \infty}{\to} 1. \qquad (7.A.80)$$

In order to evaluate F_1 over $\mathcal{B}$ I rewrite the product in terms of the gamma functions using the identity $\Gamma(z+n)/\Gamma(z) = z(z+1)\cdots(z+n-1)$ as

$$\begin{aligned} F_1(\mathcal{B}) &= \prod_{j=1}^{p} \frac{(N + (j-1)/\lambda - \kappa_j' + 1)\cdots(N + (j-1)/\lambda - q)}{(N + j/\lambda - \kappa_j')\cdots(N + j/\lambda - q - 1)} \\ &= \prod_{j=1}^{p} \frac{\Gamma(N + j/\lambda - \kappa_j')\Gamma(N + j/\lambda - q + 1 - 1/\lambda)}{\Gamma(N + j/\lambda - \kappa_j' + 1 - 1/\lambda)\Gamma(N + j/\lambda - q)}. \end{aligned} \qquad (7.A.81)$$

Using the following asymptotic relation,

$$\lim_{|z| \to \infty} \frac{\Gamma(z+a)}{\Gamma(z)} = z^a, \qquad (7.A.82)$$

I reduce $F_1(\mathcal{B})$ in Eq. (7.A.81) as

$$\begin{aligned} F_1(\mathcal{B}) &\stackrel{N \to \infty}{\to} \prod_{j=1}^{p} \left(\frac{N + j/\lambda - q}{N + j/\lambda - \kappa_j'} \right)^{1 - 1/\lambda} \\ &\to \prod_{j=1}^{p} (1 - \kappa_j'/N)^{1/\lambda - 1} \end{aligned} \qquad (7.A.83)$$

F_1 over $\mathcal{C}$, similarly, is given by

$$F_1(\mathcal{C}) \stackrel{N \to \infty}{\to} \prod_{i=1}^{q} \left(1 + \frac{1}{\lambda} \frac{\kappa_i}{N} \right)^{\lambda - 1}. \qquad (7.A.84)$$

Evaluation of $([0']_{\boldsymbol{\kappa}}^{\lambda})^2$ is straightforward and is equal to product of the following three factors

$$([0']_{\boldsymbol{\kappa}}^{\lambda})^2(\mathcal{A}) = \lambda^{-2(p-1)}\Gamma^2(p) \prod_{j=1}^{p} \frac{\Gamma^2(q - (j-1)/\lambda)}{\Gamma^2(1 - (j-1)/\lambda)}, \qquad (7.A.85)$$

$$([0']_{\boldsymbol{\kappa}}^{\lambda})^2(\mathcal{B}) = \prod_{j=1}^{p} \frac{\Gamma^2(-(j-1)/\lambda + \kappa_j')}{\Gamma^2(-(j-1)/\lambda + q)}, \qquad (7.A.86)$$

$$([0']_{\boldsymbol{\kappa}}^{\lambda})^2(\mathcal{C}) = \lambda^{-2(\sum_{i=1}^{q} \kappa_i) + 2pq} \prod_{i=1}^{q} \frac{\Gamma^2(\kappa_i - \lambda(i-1))}{\Gamma^2(p - \lambda(i-1))}. \qquad (7.A.87)$$

The value of $j_{\boldsymbol{\kappa}}^{\lambda}$ over $\mathcal{A}$ is given by

$$j_{\boldsymbol{\kappa}}^{\lambda}(\mathcal{A}) = \prod_{i=1}^{q} \prod_{j=1}^{p} (\kappa_j' + \frac{1}{\lambda}\kappa_i)^2. \qquad (7.A.88)$$

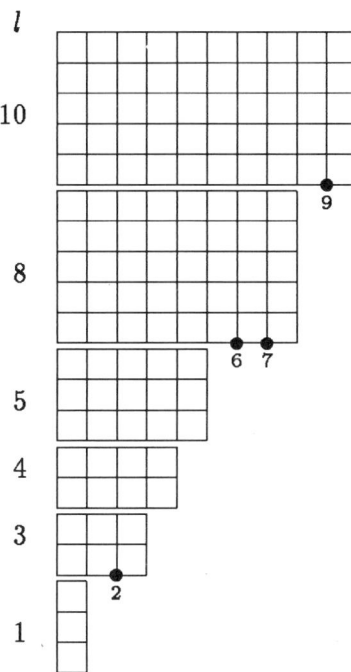

Fig. 7.9. A subdiagram $\mathcal{B}$ with $p = 10$ is divided into p *cells*. The dots ● indicate the empty cells where $\kappa'_l = \kappa'_{l+1}$, and only the 1, 3, 4, 5, 8, and 10th cells are non-empty. For the lth cell, κ_i are all same and equal to l.

In order to evaluate $j_{\boldsymbol{\kappa}}^{\lambda}$ over $\mathcal{B}$, I further divide up the subdiagram $\mathcal{B}$ into p cells such that lth cell is given by $\{(i,j),\ 1 \leq j \leq l,\ \kappa'_{l+1} + 1 \leq i \leq \kappa'_l\}$. If $\kappa'_l = \kappa'_{l+1}$, then the lth cell is empty. Fig. 7.9 illustrates how the subdiagram $\mathcal{B}$ is divided into p cells and how the empty cells appear. The value of $j_{\boldsymbol{\kappa}}^{\lambda}$ over $\mathcal{B}$ is then as follow

$$
\begin{aligned}
j_{\boldsymbol{\kappa}}^{\lambda}(\mathcal{B}) &= \prod_{l=1}^{p}\prod_{j=1}^{l}\prod_{i=\kappa'_{l+1}+1}^{\kappa'_l} (\kappa'_j - i + 1 + (l-j)/\lambda)(\kappa'_j - i + (l-j+1)/\lambda) \\
&= \prod_{l=1}^{p}\prod_{j=1}^{l} \frac{\Gamma(\kappa'_j - \kappa'_{l+1} + 1 + (l-j)/\lambda)\Gamma(\kappa'_j - \kappa'_{l+1} + (l-j+1)/\lambda)}{\Gamma(\kappa'_j - \kappa'_l + 1 + (l-j)/\lambda)\Gamma(\kappa'_j - \kappa'_l + (l-j+1)/\lambda)}
\end{aligned}
\tag{7.A.89}
$$

where $\kappa'_{p+1} \equiv q$. The contributions from the empty cells are identically equal to one. For the non-empty lth cell $\kappa_i = l$ for $\kappa'_{l+1} + 1 \leq i \leq \kappa'_l$. The expression above simplifies further to

$$
\begin{aligned}
j_{\boldsymbol{\kappa}}^{\lambda}(\mathcal{B}) &= \frac{1}{\Gamma^p(1/\lambda)} \prod_{j=1}^{p} \Gamma(\kappa'_j + 1 - j/\lambda)\Gamma(\kappa'_j - (j-1)/\lambda) \\
&\times \prod_{i>j} \frac{\Gamma(\kappa'_i - \kappa'_j + 1 - (i-j+1)/\lambda)\Gamma(\kappa'_i - \kappa'_j - (i-j)/\lambda)}{\Gamma(\kappa'_i - \kappa'_j + 1 - (i-j)/\lambda)\Gamma(\kappa'_i - \kappa'_j - (i-j-1)/\lambda)}
\end{aligned}
\tag{7.A.90}
$$

In order to take the $N \to \infty$ limit using Eq. (7.A.82), I take the following ratio first and then send the ratio to the limit,

$$
\frac{([0']_{\boldsymbol{\kappa}}^{\lambda})^2(\mathcal{B})}{j_{\boldsymbol{\kappa}}^{\lambda}(\mathcal{B})} \overset{N \to \infty}{\longrightarrow} \frac{\Gamma^p(1/\lambda)}{\prod_{j=1}^{p}\Gamma^2(q - (j-1)/\lambda)} \prod_{j=1}^{p}(\kappa'_j)^{1/\lambda - 1} \prod_{i>j} |\kappa'_i - \kappa'_j|^{2/\lambda}.
\tag{7.A.91}
$$

Similarly, $j_{\boldsymbol{\kappa}}^{\lambda}$ over $\mathcal{C}$ is given by

$$
\begin{aligned}
j_{\boldsymbol{\kappa}}^{\lambda}(\mathcal{C}) &= \frac{\lambda^{-2(\sum_{i=1}^{q}\kappa_i)+2pq}}{\Gamma^q(\lambda)} \prod_{i=1}^{q} \Gamma(\kappa_i - \lambda(i-1))\Gamma(\kappa_i + 1 - \lambda i) \\
&\times \prod_{j>i} \frac{\Gamma(\kappa_i - \kappa_j - \lambda(i-j))\Gamma(\kappa_i - \kappa_j + 1 - \lambda(i-j+1))}{\Gamma(\kappa_i - \kappa_j - \lambda(i-j-1))\Gamma(\kappa_i - \kappa_j + 1 - \lambda(i-j))}.
\end{aligned}
\tag{7.A.92}
$$

I take the following ratio and sent it to the thermodynamic limit using Eq. (7.A.82),

$$
\frac{([0']_{\boldsymbol{\kappa}}^{\lambda})^2(\mathcal{C})}{j_{\boldsymbol{\kappa}}^{\lambda}(\mathcal{C})} \overset{N \to \infty}{\longrightarrow} \frac{\Gamma^q(\lambda)}{\prod_{i=1}^{q}\Gamma^2(p - \lambda(i-1))} \prod_{j=1}^{p} \kappa_j^{\lambda-1} \prod_{i>j} |\kappa_i - \kappa_j|^{2\lambda}.
\tag{7.A.93}
$$

Putting all the terms together, setting $\kappa_j/N = x_j$ and $\kappa'_j/N = y_j$ and turning the sums into integrals I get Eq. (7.45) with the normalization constant given by Eq. (7.49).

If the differences $|\kappa'_i - \kappa'_j|$ ($|\kappa_i - \kappa_j|$) are of order $\mathcal{O}(1)$ then the corresponding contributions to Eq. (7.A.91) (Eq. (7.A.93)) are of order $\mathcal{O}(1)$ instead of $\mathcal{O}(N^{2/\lambda})$

$(\mathcal{O}(N^{2\lambda}))$ as $N \to \infty$; therefore, the contributions of such terms to the DDDCF are suppressed in the thermodynamic limit. If κ'_j is of order $\mathcal{O}(1)$ or $\mathcal{O}(N)$ then there is a corresponding contribution of order $\mathcal{O}(N^{1-1/\lambda})$; so, if $\lambda < 1$ the form factor vanishes as $N \to \infty$, and if $\lambda > 1$ it diverges. However, the net contribution of such term to the DDDCF is of order $N^{-1} \times N^{1-1/\lambda}$ which always vanishes so long as $\lambda \geq 0^{+}$. (The factor $1/N$ comes from Δy_j.) More generally, let there be r κ'_j of all order $\mathcal{O}(1)$ or $\mathcal{O}(N)$. Then, the form factor will be of order N to the power of $r(1 - 1/\lambda) - (2/\lambda)r(r-1)/2 = r(1 - r/\lambda)$. Therefore, if the $\lambda > r$ the form factor diverges. In other words such configuration is favorable. This is an exotic violation of the super selection rule since r quasiholes are stuck together. However, the net contributions of these terms are again vanishingly small except, of course, in the long-wavelength limit where only these exceptional states survive.

Almost identical method can be applied to obtain Eq. (7.55). When the limit $x, t \to 0$ is taken, the propagator becomes just the static density ρ_0. This gives a generalization of the Selberg's integral formula and might have connections with the q-deformed Lie algebra.

7.B Asymptotic Expansion

7.B.1 Dynamical density-density correlation function

A method for expanding the dynamical density-density correlation functions as $x \to \infty$ is presented in the appendix. A similar method has been used previously in [21] for the equal-time correlation functions.

The form factor to the DDDCF is largest near a phase space region where all x_j's near zero and y_i's near zero or one. Thus, the integrand will be expanded about this region. In Fig. 7.6 the shaded region gives non-zero contributions to the DDDCF and the regions with darker shades give largest contributions to the DDDCF in the long-wavelength limit. Each of these dark regions is labeled with the index n and is called nth sector, whose contribution to the correlation function has a characteristic oscillation, and is spanned by all x_i's close to zero and n $(p - n)$ y_j's close to one (zero).

To leading order in the normalized momentum variables the following relation hold true

$$Qx \pm Et = 2\pi\rho_0\xi^{\mp}\left(\sum_{i=1}^{q} x_i + \sum_{j=1}^{p-n} y_j\right) - 2\pi\rho_0\xi^{\pm}\sum_{k=1}^{n} w_k + 2\pi\rho_0 nx, \qquad (7.B.94)$$

where $\xi^{\pm} = x \mp v_s t$. The sound velocity v_s is $2\pi\rho_0\lambda$. A new set of n variables w_k for y_j close to one are introduced such that $w_k = 1 - y_j$. ξ^{+} (ξ^{-}) is the space-time coordinate for the right (left) moving particle or hole excitations.

To leading order in each of the p sectors the DDDCF is given by

$$\langle \rho(x,t)\rho(0,0)\rangle \approx \frac{C}{2}\left(I_1^\lambda(+i|q,p)\left(\frac{1}{\xi^-}\right)^2 + I_1^\lambda(-i|q,p)\left(\frac{1}{\xi^+}\right)^2\right)$$

$$+ \sum_{n=1}^{p}\tilde{C}_n\left(\frac{1}{\xi^+\xi^-}\right)^{\frac{n^2}{\lambda}}\frac{1}{2}\left(I_1^\lambda(-i|q,p-n)I_2^\lambda(+i|n)e^{i2\pi\rho_0 nx}\right.$$

$$+ \left. I_1^\lambda(i|q,p-n)I_2^\lambda(-i|n)e^{-i2\pi\rho_0 nx}\right), \tag{7.B.95}$$

where $\tilde{C}_n = \binom{p}{n}C\lambda^{-2nq}n^2\left(\frac{1}{2\pi\rho_0}\right)^{\frac{2n^2}{\lambda}-2}$ and the functions I_1^λ and I_2^λ are defined as follow

$$I_1^\lambda(z|l,p-m) = \prod_{i=1}^{l}\int_0^\infty dx_i\prod_{j=1}^{p-m}\int_0^\infty dy_j\left\{\begin{matrix}\left(\sum_{i=1}^l x_i + \sum_{j=1}^p y_j\right)^2 & \text{if } m=0\\ 1 & \text{if } m\neq 0\end{matrix}\right\}$$

$$\times \prod_{i=1}^{l}\prod_{j=1}^{p-m}(x_i+\lambda y_j)^{-2}\frac{\prod_{i<j}|x_i-x_j|^{2\lambda}\prod_{i'<j'}|y_{i'}-y_{j'}|^{2/\lambda}}{\prod_{i=1}^{l}x_i^{1-\lambda}\prod_{j=1}^{p-m}y_j^{1-1/\lambda}}$$

$$\times \exp\left(-z\left(\sum_i x_i + \sum_j y_j\right)\right) \tag{7.B.96}$$

$$I_2^\lambda(z|m) = \prod_{k=1}^{m}\int_0^\infty dw_k\frac{\prod_{i<j}|w_i-w_j|^{2/\lambda}}{\prod_{k=1}^{m}w_k^{1-1/\lambda}}\exp\left(-z\sum_k w_k\right) \tag{7.B.97}$$

Since I_1^λ and I_2^λ are absolutely convergent only if $Re(z) > 0$, it is necessary to analytically continue the functions as follow

$$I_1^\lambda(z|l,m) = \left(\frac{1}{z}\right)^{\lambda l^2 + m^2/\lambda - 2lm + 2\delta_{m,p}}I_1^\lambda(1|l,m), \tag{7.B.98}$$

$$I_2^\lambda(z|m) = \left(\frac{1}{z}\right)^{m^2/\lambda}I_2^\lambda(1|m). \tag{7.B.99}$$

The above analytical extensions give the following relations:

$$I_1^\lambda(-i|q,p-n)I_2^\lambda(+i|n) = I_1^\lambda(1|q,p-n)I_2^\lambda(1|n), \tag{7.B.100}$$
$$I_1^\lambda(\pm i|q,p) = -I_1^\lambda(1|q,p). \tag{7.B.101}$$

Finally, the DDDCF to leading order in each harmonic mode is given by

$$\langle 0|\rho(x,t)\rho(0,0)|0\rangle \approx -\frac{C}{2}I_1^\lambda(1|q,p)\left(\frac{1}{(\xi^+)^2} + \frac{1}{(\xi^-)^2}\right) + \sum_{n=1}^{p}C_n\left(\frac{1}{\xi^+\xi^-}\right)^{n^2/\lambda}\cos(2\pi\rho_0 nx),$$
$$\tag{7.B.102}$$

where $C_n = \tilde{C}_n I_1^\lambda(1|q,p-n)I_2^\lambda(1|n)$.

The renormalization parameter is defined in [38] as $\eta = 2(v_J/v_N)^{1/2}$, where v_J and v_N are the current and charge velocities. For the $U(1)$ CSM $\eta = 2/\lambda$ [31].[3] With this

---

[3]The interaction coupling constant λ used in Ref. [31] corresponds to $\lambda - 1$ in this chapter.

identification the form of Eq. (7.B.102) in the static limit agrees with the expansion given in [38]. Furthermore, by matching the universal constant in the 0th sector I deduce the following integral formula

$$I_1^\lambda(1|q,p) = (2\pi^2\lambda C)^{-1}. \tag{7.B.103}$$

7.B.2   One-particle Green's function

Using the similar method used for the DDDCF I find the following leading order terms for the hole propagator

$$\langle 0|\Psi^\dagger(x,t)\Psi(0,0)|0\rangle \approx \sum_{n=0}^{p} \mathcal{D}_n \left(\frac{1}{2\pi\rho_0\xi^+}\right)^{(n-\lambda)^2/\lambda} \left(\frac{1}{2\pi\rho_0\xi^-}\right)^{n^2/\lambda} e^{i(2\pi\rho_0(n-\lambda/2)x+(\pi\rho_0\lambda)^2t)},$$

(7.B.104)

where $\mathcal{D}_n = \rho_0 D \lambda^{1-\lambda-2(q-1)n} \binom{p}{n} I_1^\lambda(1|q-1,p-n) I_2^\lambda(1|n) e^{-i\pi(n-\lambda/2)}$. Here, $I_1^\lambda(z|q-1,p)$ is defined as before but without the pre-factor $\left(\sum_i x_i + \sum_j y_j\right)^2$.

Chapter 8

Concluding Remarks

In this book some elementary and some novel features of two families of exactly solvable models are studied. It has been observed that the strings, which are introduced originally for the NNE Bethe-ansatz-solvable Heisenberg spin chain, can be used to represent the eigenstates of the one-band Hubbard model. Furthermore, exact and complete eigenstates of the *finite size* ISE spin chain are also represented by the strings. The strings are also known to be present in the *t-J* models. It appears that the string structure is quite general and is probably a generic feature of one-dimensional integrable many-body systems and it is connected with the underlying dynamically generated quantum symmetry. The strings with length longer than one probably represent the internal quantum structures of solitonic excitations of the strongly interacting 1D systems and, therefore, may be related to non-Abelian fractional statistics which are presumably described by the quantum Yang-Baxter equations.

The ISE models have very rich and elegant structures and seem to represent idealized 1D quantum fluids supporting fractional statistics. By explicitly calculating the correlation functions and examining the excitation contents I am convinced that this class of model systems represents an emerging *fully* solvable paradigm that is a full generalization of the ideal Fermi gas. (This finite energy generalization needs to be contrasted with the Luttinger liquid which is an universal $T \to 0$ 1D fluid.) Future works should include the effects of residual interactions or gauge fluctuations on these ideal systems and examine their stabilities and instabilities.

Bibliography

[1] P.W. Anderson, Science **235**, 1196(1987).

[2] P. Kulish and E. Sklyanin, in *Integrable Quantum Field Theories*, edited by H. Araki *et al.*, Lecture Notes in Physics Vol. **151**; P. Kulish, N. Yu. Reshetikhin, and E. Sklyanin, Lett. Math. Phys. (Netherlands) **5**, 393 (1981); L. Takhtajan, Phys. Lett. **87A**, 479 (1982); G. Babudjian, Phys. Lett. **90A**, 479 (1982), and Nucl. Phys. **B215**, 317 (1983).

[3] P. A. Bares and G. Blatter, Phys. Rev. Lett. **64**, 2567 (1990).

[4] R. Baxter, Ann. Phys. **76**, 1(1973); **76**, 25(1973); **76**, 48(1973); *Exactly Solved Models in Statistical Mechanics*, Academic Press Inc.(London), 1982.

[5] A.A. Belavin, A. M. Polyakov, A. B. Zamolodchikov, Nucl. Phys. **B241**, 333(1984); P. Ginsparg in Les Houches, Session XLIX, Fields, Strings, and Critical Phenomena, 1989; J.L. Cardy in Les Houches, Session XLIX.

[6] D. Bernard, M. Gaudin, F. D. M. Haldane, and V. Pasquier, J. Phys. A:Math. Gen. **26**, 5219 (1993).

[7] D. Bernard and Y. S. Wu, (unpublished).

[8] D. Bernard, Commun. Math. Phys. **137**, 191(1991); D. Bernard, G. Felder, Nucl. Phys. **B365**, 98(1991).

[9] H. A. Bethe, Z. Phys. **71**, 265 (1931).

[10] W. Braunnck, Z. Phys. **B 28**, 291(1977).

[11] F. Calogero, J. Math. Phys. **10**, 2191 (1962); **10**, 2197(1969);

[12] J. L. Cardy, Nucl. Phys. **B270**[FS16], 186 (1986).

[13] J. des Cloizéaux, J.J. Pearson, Phys. Rev. **128**, 2131(1962).

[14] C.F. Coll, Phys. Rev. **B 9**, 2150(1974).

[15] See, for example, J. F. Cornwell, *Group theory in physics* (Academic Press, London), 1984.

[16] V. G. Drinfel'd, Sov. Math. Dokl. **32**, 254 (1985).

[17] F.J. Dyson, J. Math. Phys. **3**, 140(1962); 157 (1962); 166 (1962).

[18] K. B. Efetov, Adv. Phys. **32**, 53 (1983).

[19] F.H.L. Essler, V.E. Korepin, K. Shoutens, SUNY Stony Brook Preprint, ITP-SB-91-15.

[20] L. D. Faddeev and L. A. Takhtajan, Phys. Lett. **85A**, 375 (1981).

[21] P. J. Forrester, Phys. Lett. A **179**, 127 (1993); Nucl. Phys. B **388**, 671 (1992).

[22] P. J. Forrester, Nucl. Phys. **B**416, 377 (1994).

[23] E. Fradkin, *Field Theories of Condensed Matter Physics*, Addison-Wesley, 1991; First Ref. in [48].

[24] J. Frohlich, in *Nonperturbative Quantum Field theory*, NATO ASI Series B: Physics Vol. **185**, 71 (1987).

[25] M. Gaudin, Phys. Lett. **24A**, 55(1967); Ph.D. Thesis, University of Paris, 1967; *La fonction d'onde de Bethe*, Masson S.A.(Paris), 1983.

[26] M. Gaudin, Phys. Rev. Lett. **26**, 1301(1971); M. Takahashi, Prog. Theo. Phys. **46**, 401(1971).

[27] F. Gebhard, D. Vollhardt, Phys. Rev. Lett. **59**, 1472(1987).

[28] D. J. Gross and A. Matytsin, (unpublished); and references therein.

[29] Z. N. C. Ha, PhD Thesis (Princeton Univ.), 1992.

[30] Z. N. C. Ha, Phys. Rev. B **46**, 12205 (1992).

[31] Z. N. C. Ha and F. D. M. Haldane, Bull. Am. Phys. Soc. **37**, 646 (1992); Phys. Rev. B **46**, 9359 (1992).

[32] Z. N. C. Ha and F. D. M. Haldane, Phys. Rev. B **47**, 12459 (1993).

[33] Z. N. C. Ha and F. D. M. Haldane, Phys. Rev. Lett. **73**, 2887 (1994).

[34] Z. N. C. Ha, Phys. Rev. Lett. **73** 1574 (1994).

[35] Z. N. C. Ha, Nucl. Phys. B **435** [FS] 604 (1995).

[36] Z. N. C. Ha and F. D. M. Haldane, (unpublished).

[37] F. D. M. Haldane, J. Phys. C14 2585 (1981); A. Luther and I. Peschel, Phys. Rev. B9, 2911 (1974);

[38] F. D. M. Haldane, Phys. Rev. Lett. **47**, 1840 (1982); **48**, 569(E) (1982).

[39] F.D.M. Haldane, D.P. Arovas, unpublished.

[40] F. D. M. Haldane, Phys. Rev. Lett. **60**, 635 (1988); B. Sriram Shastry, Phys. Rev. Lett. **60**, 639 (1988).

[41] F. D. M. Haldane, unpublished.

[42] F. D. M. Haldane, Z. N. C. Ha, J. C. Talstra, D. Bernard, and V. Pasquier, Phys. Rev. Lett. **69**, 2021 (1992).

[43] F. D. M. Haldane, A. M. Tsvelik, unpublished.

[44] F. D. M. Handane, Phys. Rev. Lett. **66**, 1529 (1991).

[45] F. D. M. Haldane, Phys. Rev. Lett. **67**, 937(1991).

[46] F. D. M. Haldane, in the Proceeding of the 16th Taniguchi Symposium, Kashiko-jima, Japan, October 26-29, 1993.

[47] F. D. M. Haldane and M. R. Zirnbauer, Phys. Rev. Lett. **71**, 4055 (1993).

[48] F. D. M. Haldane, Bull. Am. Phys. Soc. **37**, 164 (1992); Phys. Rev. Lett. **67**, 937 (1991); S. Mitra and A. H. MacDonald, *ibid.* **37**, 377 (1992); P. J. Forrester and B. Jancovici, J. Phys. (Paris) **45**, L583 (1984); A. P. Polychronakos, Nucl. Phys. **B324**, 597 (1989); N. Kawakami, Phys. Rev. Lett. **71**, 275 (1993); S. Iso and S. J. Rey, (unpublished); A. D. de Veigy and S. Ouvry, Phys. Rev. Lett. **72**, 121 (1994).

[49] F. D. M. Haldane, in the proceeding of the *International Colloquium in Modern Field Theory*, Tata Institute, Bombay, India, January 5-12, 1994.

[50] B. I. Halperin, Phys. Rev. Lett. **52**, 1583 (1984).

[51] P. J. Hanlon, R. P. Stanley, and J. R. Stembridge, Contemp. Math. **138**, 151 (1992).

[52] K. Hikami and M. Wadati, J. Phys. Soc. Jpn. **62**, 1203 (1993).

[53] J. Hubbard, Proc. Roy. Soc. **A276**, 238(1963).

[54] L. Hulthén, Arkiv Mat. Astron. Fysik **26A**, No. 1(1938).

[55] V.I. Inozemtsev, J. Stat. Phys. **59**, 1143(1990).

[56] H. Jack, Proc. Roy. Soc. Edinburgh Sect. A **69** 1 (1969-1970).

[57] M. Jimbo ed., *Yang-Baxter Equation in Integrable Systems*, Advanced Series in Math. Phys. Vol. **10**, World Scientific, Singapore, 1989; T. Kohno ed., *New Developments in the Theory of Knots, ibid.* Vol. **11**; L. H. Kauffman, *Knots and Physics*, World Scientific, Singapore, 1991.

[58] L. P. Kadanoff and H. Ceva, Phys. Rev. **B3**, 3918 (1971); E. Fradkin and L. Susskind, Phys. Rev. **D17**, 2637 (1978).

[59] K. W. J. Kadell, Compos. Math. **87**, 5 (1993).

[60] V. Kalmeyer and R.B. Laughlin, Phys. Rev. Lett. **59**, 2095(1987).

[61] J. Kaneko, SIAM J. Math. Anal. **24**, 1086 (1993); K. W. J. Kadell, Adv. Math. (to be published).

[62] L.H. Kauffman, *Knots and Physics*, World Scientific, 1991; Braid Group, Knot Theory and Statistical Mechanics, Advanced Series in Mathematical Physics Vol. 9, 1989.

[63] S. Katsura, Ann. Phys. **31**, 325(1965).

[64] N. Kawakami, T. Usuki and A. Okiji, Phys. Lett. **A 137**,287(1989).

[65] N. Kawakami and T. Usuki, Phys. Rev. **B 40**, 7066(1989).

[66] N. Kawakami and S. K. Yang, Phys. Rev. Lett. **67**, 2493 (1991).

[67] N. Kawakami, Phys. Rev. **B 46**, 3191 (1992).

[68] K. von Klitzing, G. Dorda, and M. Pepper, Phys. Rev. Lett. **45** 494 (1980); D. C. Tsui, H. L. Stormer, and A. C. Gossard, Phys. Rev. Lett. **48**, 1559 (1982).

[69] K. Kubo, Prog. Theor. Phys. **64**, 758(1980).

[70] Y. Kuramoto and H. Yokoyama, Phys. Rev. Lett. **67**, 1338 (1991).

[71] M. G. G. Laidlaw and C. M. DeWitt, Phys. Rev. D **3**, 1375 (1971); D. Finkelstein and J. Rubinstein, J. Math. Phys. **9**, 1762 (1968).

[72] R. B. Laughlin, Phys. Rev. Lett. **50**, 1395 (1983).

[73] J. M. Leinaas and J. Myrheim, Nuovo Cimento B **37**, 132 (1977).

[74] A. Lenard, J. Math. Phys. **5**, 930 (1964).

[75] F. Lesage, V. Pasquier, and D. Serban, Nucl. Phys. **B435**, 585 (1995); P. J. Forrester, J. Math. Phys. **36**, 86 (1995); J. Minahan and A. P. Polychronakos, Phys. Rev. **B50**, 4236 (1994).

[76] E.H. Lieb, Phys. Rev. **162**, 162(1967); Phys. Rev. Lett. **18**, 1046(1967); Phys. Rev. Lett. **19**, 108(1967).

[77] E.H. Lieb, F.Y. Wu, Phys. Rev. Lett. **20**, 1445(1968).

[78] K.L. Liu, Can. J. Phys. **62**, 361(1984).

[79] See, for example, J. Lowenstein, in *Recent Advances in Field Theory and Statistical Mechanics* Les Houches Summer School 1982, session XXXIX, North Holland, 1984.

[80] J.M. Luttinger, J. Math. Phys. **15**, 609(1963); E. H. Lieb and D. Mattis, J. Math. Phys. **6**, 304(1965).

[81] I. G. Macdonald, Lecture Notes in Math. **1271**, 189 (1987).

[82] I. G. Macdonald, *Symmetric Functions and Hall Polynomials*, Clarendon Press (Oxford), 2nd ed., 1995.

[83] G. D. Mahan, *Many-Particle Physics*, Plenum (New York), 2nd ed., 1990.

[84] M.L. Mehta, *Random Matrices*, Academic Press (New York), 1967.

[85] W. Metzner and D. Vollhardt, Phys. Rev. Lett. **59**, 121 (1987); Phys. Rev. **B 37**, 7382 (1988).

[86] J. Moser, Adv. Math. **16**, 197 (1975).

[87] R. Orbach, Phys. Rev. **112**, 309(1958).

[88] M. Ogata, H. Shiba, Phys. Rev. **B41**, 2326(1990).

[89] A. P. Polychronakos, Phys. Rev. Lett. **69**, 703 (1992).

[90] See, for example, R. E. Prange and S. M. Girvin eds., *The Quantum Hall Effect*, Springer-Verlag, 1987; M. Stone ed., *The Quantum Hall Effect*, World Scientific, Singapore, 1992.

[91] N. Read, Phys. Rev. Lett. **65**, 1502 (1990).

[92] H. Shiba, Phys. Rev. **B 6**, 930(1972).

[93] B. D. Simons, P. A. Lee, and B. L. Altshuler, Phys. Rev. Lett. **70**, 4122 (1993).

[94] R. P. Stanley, Adv. Math. **77**, 76 (1989).

[95] B. Sutherland, J. Math. Phys. **12**, 246 (1971); **12** 251 (1971); Phys. Rev. A **4**, 2019 (1971); **5**, 1372(1972).

[96] B. Sutherland, Phys. Rev. B **12**, 3795 (1975); P. Schlottmann, Phys. Rev. B **36**, 5177 (1987); S. Sarkar, J. Phys. A:Math. Gen. **23**, L409 (1990.)

[97] See, for example, L.A. Takhtadzhan and L.D. Faddeev, Russian Math. Surveys **34**:5, 11(1979); L. A. Takhtajan, in Exactly Solvable Problems in Condensed Matter and Relativistic Field Theory, Lecture Notes in Physics **242**, Springer-Verlag, 1985.

[98] J. C. Talstra and F. D. M. Haldane, cond-math/9411065.

[99] M. Takahashi, Prog. Theor. Phys. **43**, 1619(1970).

[100] M. Takahashi, Prog. Theor. Phys. **47**, 69(1972).

[101] R. Tao and Y. S. Wu, Phys. Rev. **B30**, 1097 (1984); R. B. Laughlin, Phys. Rev. **B23**, 5632 (1981); B. I. Halperin, Phys. Rev. **B25**, 2185 (1982).

[102] D. A. Tennant *et el.*, Phys. Rev. Lett. **70**, 4003 (1993).

[103] S. Tomonaga, Prog. Theor. Phys. (Kyoto) **5**, 544 (1950).

[104] X. G. Wen, Phys. Rev. Lett. **64**, 216 (1990), Phys. Rev. B **41**, 12838 (1990); Mod. Phys. Lett. B **5**, 39 (1991).

[105] See, for example, M. Wortis, in *Chemistry and Physics of Solid Surfaces*, Vol. 7, Springer-Verlag, Berlin, 1988; N. C. Bartelt, T. L. Einstein, and E. D. Williams, Surf. Sci. Lett. **240** L591 (1990).

[106] F. Woynarovich, J. Phys. **C 15**, 85(1982).

[107] Y. S. Wu, Phys. Rev. Lett. **73**, 922 (1994).

[108] Z. Yan, Contemp. Math. **138**, 239 (1992).

[109] C.N. Yang, Phys. Rev. Lett. **19**, 1312(1967).

[110] C. N. Yang and C. P. Yang, J. Math. Phys. **10**, 1115 (1969).

[111] C.N. Yang and S. Zhang, Mod. Phys. Lett. **B4**, 759(1990).

[112] See, for example, S. C. Zhang, Int. J. Mod. Phys. **B6**, 25 (1992) and references therein.

[113] See, for example, J. Zinn-Justin, *Quantum Field Theory and Critical Phenomena*, Clarendon Press (Oxford), 2nd ed., 1993.

[114] M. R. Zirnbauer and F. D. M. Haldane, xxx e-print archive, cond-mat/9504108.

Index